Focus on Statistics

Investigations for the Integration of Statistics into Grades 9-12 Mathematics Classrooms

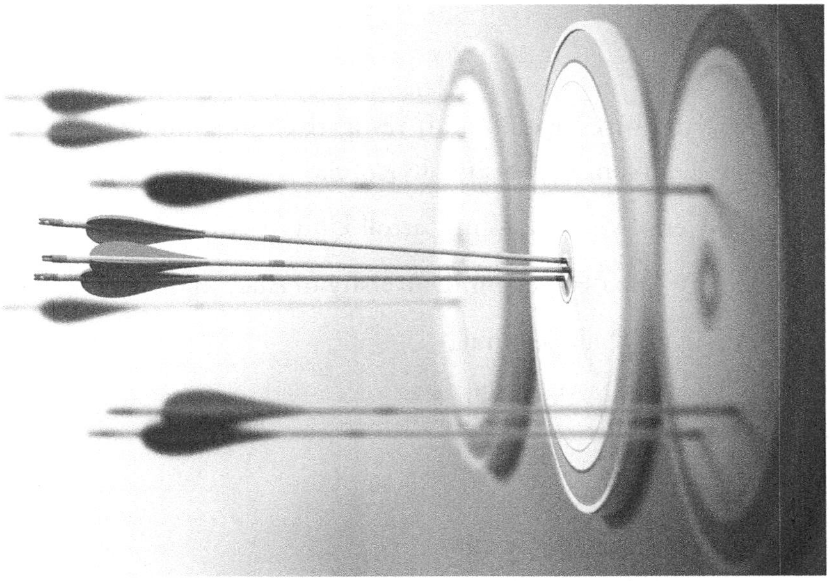

Sara Brown
Mathematics Institute of Wisconsin

Patrick Hopfensperger
Retired, University of Wisconsin-Milwaukee

Henry Kranendonk
Marquette University

Published 2020
Printed in the United States of America
10 9 8 7 6 5 4 3 2 1
ISBN: 978-1-7342235-0-7

Foreword and Investigation on Investigative Questioning by
Christine Franklin, University of Georgia

Edited by Jerry Moreno, John Carroll University

Graphics by Anna Fergusson, University of Auckland New Zealand

Cover design by Valerie Nirala

Interior design by Valerie Nirala

For information, address:
American Statistical Association
732 North Washington St.
Alexandria, VA 22314-1943

Table of Contents

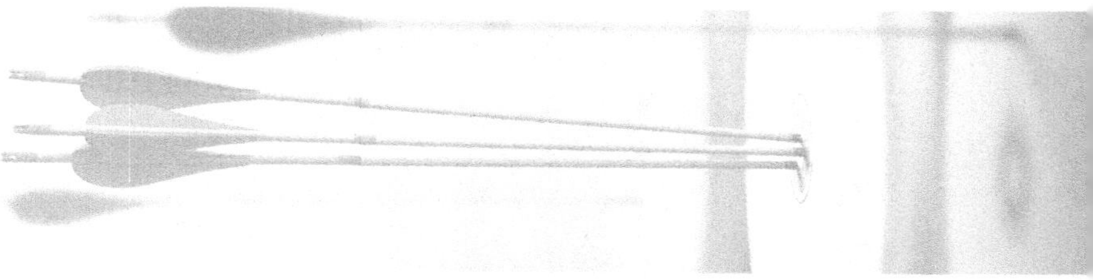

Contents

Foreword

I am honored to write the foreword for *Focus on Statistics: Investigations for the Integration of Statistics into Grades 9–12 Mathematics Classrooms,* a collection of data-centric investigations for high-school students.

Statistics is recognized as a fundamental component in the K–12 mathematics curriculum. Examinations of current state standards, many based on the Common Core State Standards (CCSS), and high-stakes national assessments such as the SAT, ACT, and National Assessment of Educational Progress (NAEP) reflect the importance of statistical literacy for all students. Continuing changes in and improvements to standards and assessments call for developing resources such as *Focus on Statistics.*

In the 2018 National Council of Teachers of Mathematics publication *Catalyzing Change in High School Mathematics,* the authors state, "Statistics and probability concepts that are essential for all high-school students support their ability to analyze data, to engage in informal statistical inference, and to understand conditional probability and independence insofar as these relate to statistical

The investigations in *Focus on Statistics* provide classroom teachers and their students with experiences to reinforce the process of statistical reasoning that is so important in making informed decisions.

thinking. Students should also leave high school with the skills necessary to be quantitatively literate, capable of reasoning with and making sense of quantitative information in order to inform the decisions that they must make now and in the future."

The investigations in *Focus on Statistics* provide classroom teachers and their students with experiences to reinforce the process of statistical reasoning that is so important in making informed decisions.

Prior to the publication of the CCSS and *Catalyzing Change in High School*

Mathematics, the American Statistical Association (ASA) produced *Guidelines for Assessment and Instruction in Statistics Education (GAISE): A Pre-K–12 Curriculum Framework,* which was approved by the ASA in 2005 (*www.amstat.org/education/gaise*). The ASA/NCTM Joint Committee on Curriculum in Statistics and Probability in 2007 worked with the GAISE Framework authors to incorporate final editing and provide funding for printing the report. GAISE 2 will be released in 2020, keeping the spirit of the original GAISE but updating with respect to advances in technology, the wealth of big data, and the importance of the statistical problem-solving process particularly related to the role of questioning in statistics.

The goals of the GAISE and GAISE 2 Framework are the following:

- Present the statistics curriculum for grades pre-K–12 as a cohesive and coherent curriculum strand (e.g., the progression of the mean from elementary to middle to secondary)
- Promote and develop statistical literacy for all students before graduating secondary school
- Provide links with the NCTM 2000 *Principles and Standards for School Mathematics* and 2018 *Catalyzing Change: Initiating Critical Conversations*
- Discuss differences between mathematical and statistical thinking, particularly the importance of context and variability within statistical thinking
- Clarify the role of probability in statistics
- Illustrate concepts associated with the statistical problem-solving process

The GAISE Framework also reinforced the need for data literacy in K–12:

- Every high-school graduate should be able to use sound statistical reasoning to intelligently cope with the requirements of citizenship, employment, and family and to be prepared for a healthy and productive life.
- Statistics education can promote the 'must-have' competencies for high-school graduates to thrive in this modern world of mass information.
- The importance of student ability to think statistically. The well-known mathematician George Polya said, "Plausible reasoning—the inferential reasoning of science and everyday life by which new knowledge is obtained—is an important part of mathematical reasoning."

The GAISE document outlines the conceptual structure for statistics education in a two-dimensional framework model with one dimension defined by

The *Guidelines for Assessment and Instruction in Statistics Education (GAISE) Report: A Pre-K–12 Curriculum Framework* was the basis for the statistics and probability conceptual category included in the Common Core State Standards in mathematics when they were released in 2010. *Focus on Statistics* follows both the GAISE Framework and the Common Core State Standards grades 9–12.

the four-step problem-solving process (formulate questions, collect data, analyze data, and interpret results), plus the nature of variability. The second dimension is comprised of three levels of statistical development (Levels A, B, and C) that students must progress through to develop statistical understanding. Grade ranges for attainment of each level are intentionally unspecified. Students must begin and master the concepts at Level A before moving on to Levels B and C. It is paramount for students to have worthwhile experiences at Levels A and B during their elementary school years to prepare for development at Level C at the secondary level. Without such experiences, a middle- (or high-) school student who has had no experience with statistics will need to begin with Level A concepts and activities before moving to Level B.

The GAISE Framework has become an instrumental document in providing guidance to writers of national mathematics documents, state standards, and assessment items; curriculum directors; pre-K–12 teachers; and faculty of teacher preparation colleges on the essential topics and concepts in data analysis and probability for all students as they progress from kindergarten to graduation from high school.

The GAISE Framework has influenced the statistics components of both the Mathematics and Statistics College Board Standards for College Success (2007) and the NCTM document *Focus on High School Mathematics* (2008). The GAISE Framework also influenced the statistics and probability strand of many state mathematics standard revisions (which includes my home state of Georgia). The GAISE Framework was the basis for the statistics and probability conceptual category included in the Common Core State Standards in mathematics when these standards were released in 2010. The GAISE Framework also influenced the LOCUS (Levels of Conceptual Understanding in Statistics), an NSF-funded project that developed assessments to measure students' understanding across levels of development as identified in GAISE. GAISE also influenced the NCTM's *Catalyzing Change in High School Mathematics*. Internationally, the GAISE Framework has been influential. There is now a Spanish version of the GAISE Framework.

Focus on Statistics is an excellent classroom resource that follows both the GAISE Framework and the Common Core State Standards grades 9–12. High school is critical in providing investigations that will further the skills needed for our students to grow and evolve into sound statistical thinkers. The investigations bring the real world to the student and provide the student the opportunity to understand the necessity of statistical reasoning and sense making for everyday life and post-secondary education.

I'm appreciative to the writers of *Focus on Statistics* and the ASA/NCTM Joint Committee for developing this valuable resource in support of the recommendations of GAISE, the recommendations of the Common Core State Standards, *Catalyzing Change in High School Mathematics*, and statistical reasoning in our high-school curriculum.

Christine Franklin
ASA K–12 Statistical Ambassador

Acknowledgments

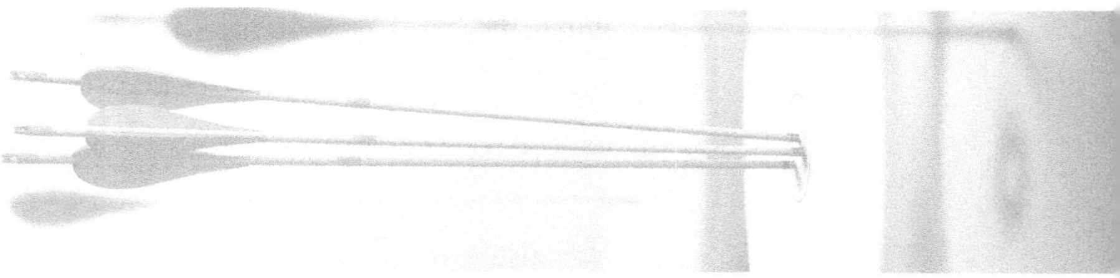

We are indebted to the ASA/NCTM Joint Committee on Curriculum in Statistics and Probability for its support throughout the process of creating and publishing *Focus on Statistics*. This project is a continuing effort of the Joint Committee to provide classroom-ready investigations written in the spirit of the GAISE framework.

The ASA has been involved in numerous publications, including *Making Sense of Statistical Studies* (15 high-school activities on surveys, observational studies, and experiments) and *Bridging the Gap* (*BTG*), which focused on statistical investigations for grades K–8.

In late 2016, the Joint Committee approved Sara Brown, Pat Hopfensperger, and Henry Kranendonk as the main writers of *Focus on Statistics*. This publication expands on the activities presented in *BTG* and activities found in the Data-Driven Mathematics (DDM) series and Quantitative Literacy Series (QL).

Sincere thanks are extended to Christine Franklin and Anna Bargagoliotti for their support expressed in the foreword and for their section on clarifying what constitutes an investigative question.

> This publication expands on the activities presented in *Bridging the Gap* and activities found in the Data-Driven Mathematics series and Quantitative Literacy Series.

Each investigation was reviewed by three high-school teachers and two statistics educators. We are thankful for their excellent comments and suggestions, which improved our writing significantly. The reviewers included Jerry Moreno of John Carroll University; Anna Fergusson of the University of Auckland, New Zealand; Alex Blohm of Loyola University Chicago; Melissa Hongsermeier of South Milwaukee High School; and Patricia Talarczyk of Mentor High School, Mentor, Ohio.

We are deeply in debt to Jerry Moreno, who was our main editor. He spent countless hours editing and giving excellent suggestions for improving each investigation.

We also extend our gratitude to Anna Fergusson, who did a remarkable job producing graphs of publishable quality for *Focus on Statistics* and providing deep dedication to and support of Joint Committee efforts.

We give special thanks to ASA Editor and Content Strategist Valerie Nirala, whose editorial and design magic brought life to our writings, without which much of what we had to offer would have lacked reader appeal.

And finally, a note of appreciation to ASA Director of Education Rebecca Nichols, whose leadership and direction were much appreciated in helping us achieve our publication goals.

Sara Brown
Pat Hopfensperger
Henry Kranendonk

About Focus on Statistics

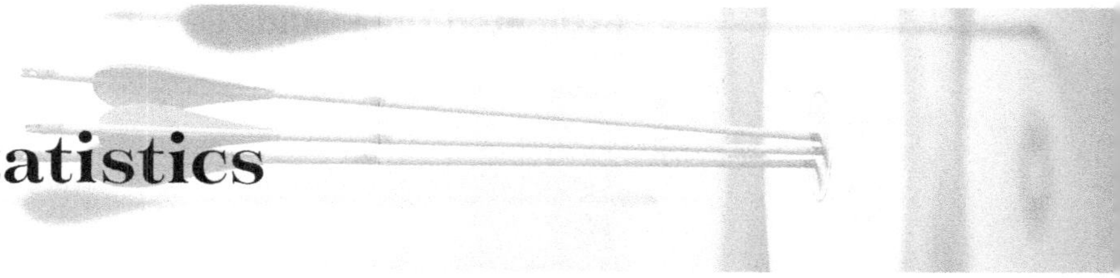

Focus on Statistics consists of 19 investigations in statistics for grades 9–12. It is written to help classroom teachers implement key statistical concepts in their classrooms. Each investigation consists of the following headings appropriately written for its specific content:

- Overview (including suggested GAISE level)
- Learning Goals
- Mathematical Practices Through a Statistical Lens
- Materials
- Estimated Time
- Pre-Knowledge
- Instructional Plan (consisting of a brief overview and the four steps of the GAISE process)
- Exit Ticket
- Extensions and Additional Ideas

The 19 investigations are separated into the following topic sections:

- **Section 1:** Getting Started
  Includes one investigation on questioning through the investigative process
- **Section 2:** One-Variable Data Analysis
  Includes four investigations, 1–4
- **Section 3:** Two-Variable Data Analysis
  Includes seven investigations, 5–11
- **Section 4:** Probability
  Includes three investigations, 12–14
- **Section 5:** Inference
  Includes four investigations, 15–18
- **Section 6:** Teacher Resources
  Includes overview of ASA online resources

Included in the 19 investigations are three exploratory investigations that encourage students to collect and analyze their own data. In Investigation 4, students collect data from the Census at School website involving the amount of homework for different grade levels. In Investigation 8, students collect data on two quantitative variables involving the length of time for dominoes to fall. In Investigation 18, students use the Census at School website to collect and compare data pertaining to stress levels for students in the US and New Zealand.

In addition, reference to various standards from "Mathematical Practices

Estimated Time Guide

Investigation	Estimated Class Periods (50-minute class period)
Questioning and statistical problem-solving process	1
1. Graphical Displays	1–2
2. Comparing Multiple Groups	1–2
3. Standard Deviation	2
4. Exploratory - Homework	2
5. Scatterplots	1–2
6. Correlation	2
7. Residual Plots	2
8. Exploratory - Dominoes	2
9. Taking a Survey	1
10. Two-Way Tables	2
11. Independent Events	2
12. Simulation	1
13. Expected Value	1
14. Normal Distribution	1–2
15. Sampling Distribution	1–2
16. Testing a Claim	1
17. Bootstrapping	1
18. Exploratory - Difference Between Two Proportions	2

Through a Statistical Lens" is made. Each investigation explicitly contains the four components of the problem-solving process presented in the American Statistical Association's *Guidelines for Assessment and Instruction in Statistics Education (GAISE) Report: A Pre-K–12 Curriculum Framework* (*www.amstat.org/education/gaise*). The GAISE framework emphasizes hands-on learning of statistics by using four steps: formulating a statistical question that can be answered with data; designing and implementing a plan to collect appropriate data; analyzing the collected data by graphical and numerical methods; and interpreting the results of the analysis in the context of the original question. A second component of the GAISE framework is comprised of three levels of statistical development:

A, B, and C. These levels are independent of age and grade level. Students progress through these levels as they continue to have more experiences with concepts of statistics and probability.

Each investigation encourages student involvement through the use of student worksheets. These worksheets provide guidance as students, working in groups, follow the four statistical problem-solving steps. The worksheets are available as a word document at *www.statisticsteacher.org/statistics-teacher-publications/focus*.

A brief overview of each section follows:

Section 1: Getting Started
This investigation introduces the different types of questions used throughout the statistical problem-solving process. The primary focus is exploring what constitutes a well-written investigative question—one that can be answered with data.

Section 2: One-Variable Data Analysis
In this section, dot plots and box plots are used to display data. The standard deviation is explored as a measure of variation.

Section 3: Two-Variable Data Analysis
In the first part of this section, the concepts of correlation and the least squares regression line are developed. In the second part, the concept of association between two categorical variables is developed. In addition, the concept of independent events is investigated.

Section 4: Probability
The GAISE framework views probability as a mathematical model and a tool for statistics. This section develops the concept of a probability distribution, expected value, and the normal distribution.

Section 5: Inference
This section develops the concept of a sampling distribution of sample proportions, testing a claim about a proportion, and using the bootstrapping method to develop a confidence interval.

Section 6: Resources
This section contains teacher resources available on the American Statistical Association website. Statistics Education Web, Statistics Teacher Network, Census at School, statistics education webinars, and publications pertaining to K–12 education are included.

Focus on Statistics is designed so each lesson can stand alone. Our goal is to provide you with a resource to give you and your students data analysis experiences that bring the essential concepts in statistics to life. It also is designed to give you flexibility. Several investigations can be completed in one to two 50-minute class periods. Many investigations suggest students collect data; if this data collection is done during class time, then additional class time is needed. However, the data collection could be completed outside of class time, as well. For planning purposes, the Estimated Time Guide shows an estimate of the time required to complete each investigation. Exit tickets are provided and could be used at the end of the lesson for formative assessment and/or extra practice. Extension activities would require additional class time.

Linking Investigations

GAISE, CCSS, Statistical Lens, and Catalyzing Change

Linking GAISE Levels and Focus on Statistics

Investigation	Level A	Level B	Level C
Questioning and Statistical Problem-Solving Process		x	
1. Graphical Displays	x		
2. Comparing Multiple Groups	x	x	
3. Standard Deviation		x	
4. Exploratory Lesson		x	
5. Scatterplots	x		
6. Correlation		x	
7. Assessing Linear Fit		x	
8. Exploratory Lesson		x	
9. Analyzing Categorical Data		x	
10. Summarizing Bivariate Categorical Data		x	
11. Comparing Conditional Relative Frequencies		x	x
12. Simulation		x	
13. Expected Value		x	
14. Normal Distribution			x
15. Sampling Distribution			x
16. Testing a Claim			x
17. Bootstrapping			x
18. Exploratory Lesson			x

Linking Grade Levels and Common Core State Standards

Investigation	Grades 6–8	High School
Questioning and Statistical Problem-Solving Process	6.SP.A.1	
1. Graphical Displays	6.SP.A.2, 6.SP.A.3, 6.SP.B.4, 6.SP.B.5.C, 6.SP.B.5.D	HSS.ID. A.1, HSS.ID. A.2, HSS. ID. A.3
2. Comparing Multiple Groups	7.SP.B.3, 7.SP.B.4	HSS.ID. A.1, HSS.ID. A.2, HSS. ID. A.3
3. Standard Deviation		HSS.ID. A.2
4. Exploratory Lesson		Review of Standards from Lessons 1 to 3
5. Scatterplots	8.SP.A.1	HSS.ID.B.6
6. Correlation	8.SP.A.1, 8.SP.A.2	HSS.ID.B.6, HSS.ID.C.8
7. Assessing Linear Fit	8.SP.A.3	HSS.ID.B.6.B, HSSS.ID.C.7
8. Exploratory Lesson		Review of Standards from Lessons 5 to 7
9. Analyzing Categorical Data	8.SP.A.4	HSS.IC.B.3, HSSS.IC.A.1
10. Summarizing Bivariate Categorical Data	8.SP.A.4	HSS.ID.B.5
11. Comparing Conditional Relative Frequencies		HSS.CP.A.4, HSS.CP.A.5
12. Simulation	7.SP.C.8.C	HSS.IC.A.2
13. Expected Value		HSS.MD.A.2, HSS.MD.A.4
14. Normal Distribution		HSS.ID.A.4
15. Sampling Distribution	7.SP.A.2	HSS.IC.A.1
16. Testing a Claim		HSS.IC.A.2
17. Bootstrapping		HSS.IC.B.4
18. Exploratory Lesson		HSS.IC.B.5

The Statistical Education of Teachers (SET) report outlines the content and conceptual understanding teachers need to know when assisting their students develop statistical reasoning skills. SET is intended for everyone involved in the statistical education of teachers, both the initial preparation of prospective teachers and the professional development of practicing teachers. PDF download at *www.amstat.org/asa/files/pdfs/EDU-SET.pdf*.

Linking Investigations to Mathematical Practices Through a Statistical Lens

Investigation	Mathematical Practices
Questioning and the Statistical Problem-Solving Process	MP 2. Reason abstractly and quantitatively
1. Graphical Displays	MP 2. Reason abstractly and quantitatively
2. Comparing Multiple Groups	MP 6. Attend to precision
3. Standard Deviation	MP 7. Look for and make use of structure
4. Exploratory Lesson	MP 1. Make sense of problems and persevere in solving them
5. Scatterplots	MP 3. Construct viable arguments and critique the reasoning of others
6. Correlation	MP 7. Look for and make use of structure
7. Assessing Linear Fit	MP 7. Look for and make use of structure
8. Exploratory Lesson	MP 1. Make sense of problems and persevere in solving them
9. Analyzing Categorical Data	MP 2. Reason abstractly and quantitatively
10. Summarizing Bivariate Categorical Data	MP 2. Reason abstractly and quantitatively
11. Comparing Conditional Relative Frequencies	MP 2. Reason abstractly and quantitatively
12. Simulation	MP 5. Use appropriate tools strategically
13. Expected Value	MP 4. Model with mathematics
14. Normal Distribution	MP 4. Model with mathematics
15. Sampling Distribution	MP 8. Look for and express regularity in repeated reasoning
16. Testing a Claim	MP 3. Construct viable arguments and critique the reasoning of others
17. Bootstrapping	MP 5. Use appropriate tools strategically
18. Exploratory Lesson	MP 1. Make sense of problems and persevere in solving them

The NCTM's *Catalyzing Change in High School Mathematics* provides the essential concepts in statistics and probability necessary for all high-school students to be statistically literate. Listed on the following page are the four statistics and probability focus areas presented in *Catalyzing Change in High School Mathematics* and the investigations whose goals match the concepts listed under each focus area.

Linking Investigations to NCTM's Catalyzing Change in High School Mathematics – Essential Concepts in Statistics and Probability

Essential Concepts	Investigation
Focus 1: Essential Concepts in Quantitative Literacy	
Making and defending informed data-based decisions is a characteristic of a quantitatively literate person.	4, 8 and 18
Focus 2: Visualizing and Summarizing Data	
Distributions of quantitative data in one variable should be described in the context of the data with respect to what is typical (the shape, with appropriate measures of center and variability, including standard deviation) and what is not (outliers), and these characteristics can be used to compare two or more subgroups with respect to a variable.	1, 2, 3, 4, and 14
The association between two categorical variables is typically represented by using two-way tables and segmented bar graphs.	9 and 10
Scatterplots can reveal patterns, trends, clusters, and gaps that are useful in analyzing association between two contextual variables.	5
Analyzing the association between two quantitative variables should involve statistical procedures, such as examining (with technology) the sum of squared deviations in fitting a linear model, analyzing residuals for patterns, generating a least-squares regression line and finding a correlation coefficient, and differentiating between correlation and causation.	6, 7, and 8
Focus 3: Statistical Inference	
The larger the sample size, the less the expected variability in the sampling distribution of a sample statistic.	15
The sampling distribution of a sample statistic formed from repeated samples for a given sample size drawn from a population can be used to identify typical behavior for that statistic.	15, 16, and 17
Focus 4: Probability	
Two events are independent if the occurrence of one event does not affect the probability of the other event. Determining whether two events are independent can be used for finding and understanding probabilities.	11 and 18
Conditional probabilities—that is, probabilities that are "conditioned" by some known information—can be computed from data organized in contingency tables.	10 and 11

Linking Investigations
Standard High-School Mathematics Curriculum

Section 1: Getting Started
Investigation on Questioning Through an Investigative Process

This investigation should be used in a first-year high-school course that integrates topics of mathematics and statistics before starting a unit on statistics. The focus is on the four-step statistical problem-solving process and the role of questioning throughout. Criteria for identifying and writing an investigative question are introduced. Each of the following 18 investigations is designed around an investigative question. As a result, students' understanding of what constitutes an investigative question is important as a launch for each investigation.

Section 2: One-Variable Data Analysis
Investigations 1–4

These four investigations may be used to teach or review a student's understanding of the skills and concepts used in communicating with data. The development of the investigation process helps students understand how data can be used to answer questions or present convincing arguments. The investigations may be used after students have studied ratio, proportion, and percent or as part of a beginning unit on statistics. These investigations are also appropriate to use when students are beginning to formalize their work with variables. The investigations could be used as a first or second unit in a first-year high-school course that integrates topics of mathematics and statistics.

Section 3: Two-Variable Data Analysis
Investigations 5–11

These investigations are about graphing and assessing the fit of linear functions to bivariate data. The investigations can be used in a first-year high-school course that integrates topics of mathematics and statistics in a variety of ways, most effectively when integrated into the unit on writing the equation of a line. They can also be used after students have completed a section on solving equations in one variable to illustrate how to apply those concepts in real-world contexts and provide investigations into graphical representations of linear relationships.

First-Year High-School Course Integrating Topics of Mathematics and Statistics

Topic	Investigations
Introduction to Statistics	1–4
Linear Functions	5–7 or 5–8
Statistics Unit or after Linear Functions	9 and 10

Investigations 6 to 8 can also be used in a second-year class that expands on the topics of mathematics and statistics when integrated into the unit on the study of functions.

Investigations 9 and 10 can be used in a first-year high-school course that integrates topics of mathematics and statistics after students have studied ratio, proportion, and percent or as part of a unit on statistics.

Investigations 10 and 11 can also be used in conjunction with a probability unit.

Section 4: Probability Investigations 12–14

The concepts on random variables, probability distributions, and expected values require subtle reasoning. It is recommended that these investigations—along with investigations 15–18—be used in a second-year course that expands on the topics of mathematics and statistics, or at least no earlier than late in a first-year high-school course that integrates topics

Second-Year High-School Course Integrating Topics of Mathematics and Statistics

Topic	Investigations
Linear Functions	5–8 or 6–8
Statistics Unit or after Linear Functions	9–11 or 10 and 11
Probability Unit	12–14
Probability Unit after Investigations 11–12	15–18

of mathematics and statistics. This is more for the level of reasoning required than for any particular set of algebraic skills needed to do the work.

These three investigations can be used in a second-year course that expands on the topics of mathematics and statistics in conjunction with a probability unit. Investigation 14 can be used after students have studied functions.

Section 5: Inference Investigations 15–18

These four investigations can be used in a second-year high-school course after completing lessons on simulations (Investigation 12) and independent events (Investigation 11). The investigations can be part of a probability unit in a second-year course that expands on the topics of mathematics and statistics or an advanced math course. Again, the level of reasoning required is the reason these lessons are more suited to upper-level courses.

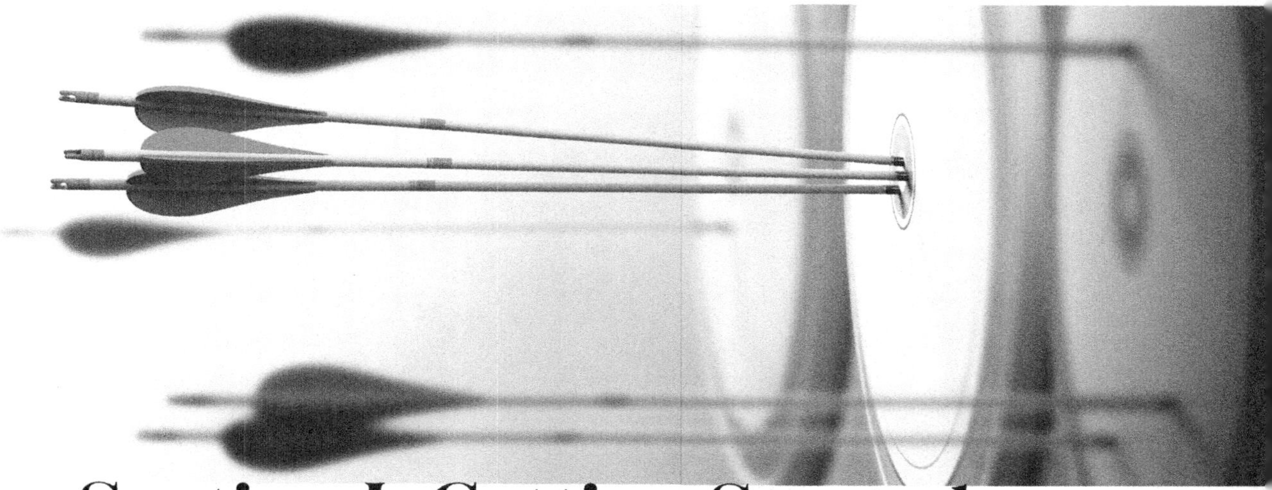

Section I: Getting Started

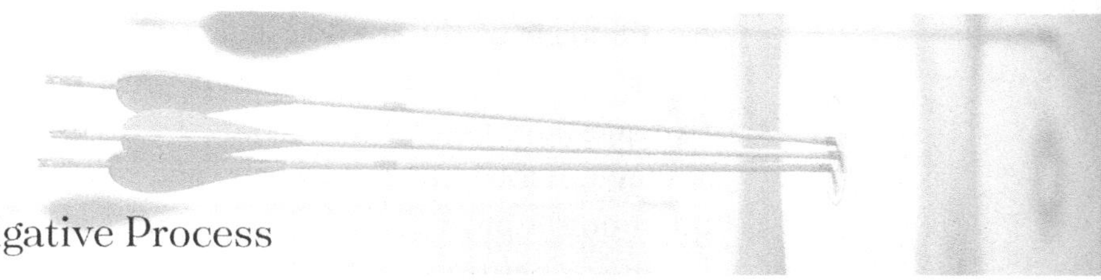

Investigation

Questioning Through the Investigative Process

Overview

As described in the foreword, statistics is recognized as a necessary component in the K–12 mathematics curriculum that is reflected in current state standards, many based on the Common Core State Standards (CCSS) and high-stakes national assessments such as SAT, ACT, and NAEP.

To support and further elaborate on the Statistics and Probability standards, the American Statistical Association (ASA) produced *Guidelines for Assessment and Instruction in Statistics Education (GAISE): A Pre-K–12 Curriculum Framework,* which was approved by the ASA in 2005 (*www.amstat.org/education/gaise*). The ASA/NCTM Joint Committee on Curriculum in Statistics and Probability in 2007 worked with the authors of the GAISE Framework to incorporate final editing and provide funding for printing the report in book format. GAISE 2 will be released in 2020, keeping the spirit of the original GAISE but updating with respect to advances in technology, the wealth of big data, and the importance of the statistical problem-solving process particularly related to the role of questioning in statistics.

Goals of the GAISE and GAISE 2 Framework are the following:

- Present the statistics curriculum for grades Pre-K–12 as a cohesive and coherent curriculum strand (e.g., the progression of the mean from elementary to middle to secondary)
- Promote and develop statistical literacy for all students before graduating from secondary school
- Provide links with the NCTM 2000 *Principles and Standards for School Mathematics* and 2018 *Catalyzing Change: Initiating Critical Conversations*
- Discuss differences between mathematical and statistical thinking, particularly the importance of context and variability within statistical thinking
- Clarify the role of probability in statistics
- Illustrate concepts associated with the statistical problem-solving process

The framework stresses hands-on active learning and emphasizes that statistical analysis is an investigative process that turns loosely formed ideas into scientific studies by doing the following:

- Formulating a question that can be answered with data
- Designing a plan to collect appropriate data
- Analyzing the collected data utilizing graphical and numerical methods
- Interpreting the results to reflect insight on the original question

The investigative process requires the investigator to formulate questions throughout the statistical problem-solving steps that will be the focus of the study. There are **research** questions that motivate the study, **investigative** questions that can be answered with data, **survey** questions to collect the data, **analysis** questions to prompt which graphs and calculations to perform, and **interpretation** questions to help focus the drawing of conclusions.

The purpose of this lesson is to help your students learn how to question throughout the statistical problem-solving process and learn how to formulate an investigative question—a question that can be answered with data.

Note: The Common Core State Standards for Mathematics (CCSSM) uses the vocabulary

Learning Goal

Understand how to use questioning throughout the statistical problem-solving process and how to construct a good investigative question.

Mathematical Practices Through a Statistical Lens

MP1. Make sense of problems and persevere in solving them.

Statistically proficient students understand how to carry out the four steps of the statistical problem-solving process.

MP2. Reason abstractly and quantitatively.

Statistically proficient students reason abstractly about the investigative problem at hand, understanding this requires the clarity of the variables that need to be measured in the data-collection process.

MP6. Attend to precision.

Statistically proficient students understand needing to be precise about the words used to ensure the intent of the investigative question is clear, needing to name variables and populations correctly.

Materials

Student worksheets are available at *www.statisticsteacher.org/statistics-teacher-publications/focus.*

» Student Worksheet 1: Questioning Throughout the Investigative Process

» Student Worksheet 2: Investigative Questions

» Student Worksheet 3 (Optional): Investigative Process

» Exit Ticket

Estimated Time

One 50-minute class period

"statistical question." Standard 6.SP.1 states, "Recognize a statistical question as one that anticipates variability in the data related to the question and accounts for it in the answers." The GAISE 2 Report, to be published in 2020, uses the term investigative question rather than statistical question. Whichever term is used, questions to analyze need to be answered with data. Throughout the investigations in *Focus on Statistics*, the term statistical question has been used.

Instructional Plan

Explain to your students that statistics is an investigative process guided by the following four statistical problem-solving steps:

1. Formulating a question that can be answered with data
2. Designing a plan to collect appropriate data
3. Analyzing the collected data utilizing graphical and numerical methods
4. Interpreting the results to reflect insight on the original question

Note: You may want to post these four steps.

Discuss that the statistical investigative process requires the investigator to use questions throughout the four steps. There are **research** questions that motivate the study, **investigative** questions that can be answered with data, **survey** questions to collect the data, **analysis** questions to prompt which representations such as graphs to construct and numerical methods to perform, and **interpretation** questions to help focus the drawing of conclusions.

Hand out Student Worksheet 1: Questioning Throughout the Investigative Process. Ask your students to read the summary, and then discuss the different types and examples of questions—investigative questions, survey questions, and analysis questions.

Student Worksheet 1: Questioning Throughout the Investigative Process

Administrators and teachers within a large school district are concerned about the perceived lack of sleep among middle-school students. Students more than ever are seemingly tired in class and struggling to stay focused. There have been complaints from parents that too much homework is being assigned. The school district believes there are several potential factors contributing to sleep deprivation among the students that are not academic. These include the number of extracurricular activities middle-school students are actively involved in and time spent on the internet and electronic devices. The school district decides to conduct a survey of selected middle-school students in the large district to investigate potential factors contributing to sleep deprivation among their middle-school students.

Formulate a question that can be answered with data. A possible *investigative question* that will assist the school district with this scenario is: What are the number of extracurricular activities middle-school students in the school district are actively involved in throughout the school year?

Collect data. A possible *survey question* for gathering data to help answer the investigative question is: How many extracurricular activities do you actively participate in during the school year?

Note: At this stage, ask your students whether the data will assist in answering the investigative question and whether there are other survey questions that need to be asked.

Analyze the data. Possible *analysis questions* are: What is an appropriate graphical display to show the distribution for the number of extracurricular activities for middle-school students? What is the shape of the distribution of the number of extracurricular activities? How much does the number of extracurricular activities vary in the distribution? What is a typical interval for the number of extracurricular activities, and where does the number of extracurricular activities cluster? What is the mean number of extracurricular activities? Are there unusual values?

Interpret the results. Ask if the analysis makes sense within the context of the situation and investigative question. Connect the results of the analysis questions to the context of the posed investigative question and make a conclusion.

Explain to your students that we now want to elaborate on how to recognize and write good investigative questions—a question that can be answered with data.

Ask your students what might be important to have in an investigative question—a question that can be answered with data.

Collate their ideas on the board, testing them as you go against the criteria for what makes a good investigative question.

Prompts you might use include the following:

» What data are the question about? – Leading to the variable(s) of interest needing to be clear in the investigative question

» Who is the question about? – Leading to the population or group of interest needing to be clear in the investigative question

» What sort of analysis does the question suggest we do with the data? – Leading to the intent of the investigative question being clear (Is it wanting to investigate summarizing data for one variable, comparing two or more groups with respect to one variable, or looking at the association between two variables?)

» What data will we collect to answer the data? – Leading to the data is/will be available to answer the investigative question (or if we are using secondary data, the investigative question can be answered with the data we have).

» Is the investigative question useful? Does it have a purpose? Is it interesting?

» Does the investigative question allow for statistical analysis to the whole group you are interested in studying?

Share with your students that criteria to consider for a well-written investigative question are the following:

» The investigative question clearly states the variable(s) of interest.

» The investigative question clearly states the population of interest.

» The investigative question clearly states the intent (summary, comparison, or association investigation).

» It should be clear from the investigative question whether the data (measurements) can be collected (called primary data) to help answer the question or if the data are already available (called secondary data).

» The investigative question is worth investigating, it is interesting and/or it has a purpose, and it assists in answering the research question.

» The investigative question allows for statistical analysis to be made of the whole group.

Refer to the investigative question listed in Step 1 on Student Worksheet 1 Questioning Throughout the Investigative Process.

1. What is the number of extracurricular activities middle-school students in the coastal school district are actively involved in throughout the school year?

Ask the students to consider the criteria for what makes a good investigative question. They will also need to decide what data need to be collected.

Answers:

Considering the criteria

1. *Variable of interest: The number of extracurricular activities students are actively involved in throughout the school year*
2. *Population: Middle-school students in the school district*
3. *Intent is clear: This will be a summary or descriptive analysis for one variable.*
4. *Data: The data are the number of extracurricular activities the sampled students actively participate in during the school year.*
5. *Interesting/purposeful: This area of the investigation is of interest to the school district, especially if they want to provide evidence that it is not (or not only) homework contributing to students being tired and unable to focus.*
6. *About the whole group: This question considers the whole group of school district middle-school students.*

Pose a second example investigative question for your students to consider.

Example: For high-school district students, do algebra students who complete their homework tend to score better on an algebra test than algebra students who do not complete their homework?

Ask the students to check this question against the criteria for what makes a good investigative question. They will also need to decide what data need to be collected.

Answers:

Considering the criteria

1. *Variable of interest: The typical test score on an algebra test*
2. *Population: School district algebra students who complete their homework and school district algebra students who do not complete their homework*
3. *Intent is clear: This will be a comparative study of two groups' (one group completes homework and one group does not) test scores.*
4. *Data: For the sampled students, test scores on an algebra test and whether the students do their homework*
5. *Interesting/purposeful: The area of investigation is of interest to the school district, especially if they want to provide evidence that doing homework improves algebra test scores.*
6. *About the whole group: This question considers the whole of both groups of school district algebra students.*

The data to be collected are test scores on an algebra test and whether the students do their homework.

Pose a third example investigative question for your students to consider. This example is not well written. Through considering the criteria, identify what is missing and how the investigative question could be improved.

Question: Are members of the boys' soccer team fitter than the members of the boys' football team?

Ask the students to consider the criteria for what makes a good investigative question. They will also need to decide what data need to be collected.

Answers:

Considering the criteria

1. *Variable of interest: Fitter. It is not clear what fitter means; the variable needs to be redefined or improved.*
2. *Population: Boys' soccer team and boys' football team, but we don't know if this is from one school or if it is primary, middle, or secondary school. The population needs to reflect the actual population from which data will be collected (e.g., the 8th-grade boys' soccer team and the 8th-grade boys' football team).*
3. *Intent is clear: This will be a comparative analysis, though the investigative question suggests the boys' soccer team would all be fitter than the boys' football team. To improve, use the idea of tendency (e.g., Do the members of the boys' soccer team tend to be fitter than the members of the boys' football team?).*
4. *Data: How fitter will be measured needs to be determined first. It is not clear what data need to be collected from the investigative question.*
5. *Interesting/purposeful: The area of investigation is of interest to the coaches and athletes, especially if they want to use the information to improve coaching and playing techniques.*
6. *About the whole group: Once the population is more clearly specified, this question can consider the whole of both groups.*

The data to be collected would be based on the criteria that are decided to judge fitness.

Ask students if this third example investigative question helps explain why middle-school students may lack a reasonable amount of sleep—the problem being investigated by the coastal school district.

Possible answer: *It is not clear how understanding whether the fitness of boys' soccer players is better than the fitness of boys' football players may or may not contribute to a lack of sleep.*

Hand out Student Worksheet 2: Investigative Questions. Place students into groups of three and ask them to complete Part I of the worksheet.

After students have completed Part I, review the answers. For the questions needing improvement, discuss possible investigative questions, and then ask your students to complete Part II.

Possible answers for Student Worksheet 2 are shown on the following pages.

Student Worksheet 2: Investigative Questions

Part I: For each investigative question, consider the criteria for a well-written investigative question. Give your reason for deciding whether the investigative question is well written or needs improvement.

Question	Criteria to Consider	Explain What Criteria the Question Does Not Meet, If Any
Are cars speeding in the posted school zone?	*1. Variable of interest is clear:* *2. Population is clear:* *3. Intent is clear:* *4. Data:* *5. Interesting/purposeful:* *6. About the whole group:*	*'Speeding' needs to be defined; what school zone needs to be specified and time of day; is the intent to summarize the proportion of cars speeding or a typical speed over the speed limit?*
What proportion of seniors at the high school participate in school-sponsored activities that take place after the school day?	*1. Variable of interest is clear:* *2. Population is clear:* *3. Intent is clear:* *4. Data:* *5. Interesting/purposeful:* *6. About the whole group:*	*The population needs to be more specific—what high school? Note: Since proportion is specified in question as the summative value desired, the variable of interest is categorical: yes or no as to whether the student participates. As written, it is not clear this question is about the whole group, as it is only asking about those senior students who participate in school-sponsored activities (one category of the categorical variable). A better question might be to ask about what activities all senior-high school students participate in. Include in the question school- and non-school-sponsored, and then the current given question would be an analysis question.*
How much money is spent on the daily lunch program?	*1. Variable of interest is clear:* *2. Population is clear:* *3. Intent is clear:* *4. Data:* *5. Interesting/purposeful:* *6. About the whole group:*	*Population is not clear—elementary, middle, or high school, and where is the school located? What time period—per day, per month, per year?*

Table continued on next page

Table continued

Question	Criteria to Consider	Explain What Criteria the Question Does Not Meet, If Any
Does taking 500 mg of Vitamin C daily protect high-school students from catching a cold during the winter months?	*1. Variable of interest is clear:* *2. Population is clear:* *3. Intent is clear:* *4. Data:* *5. Interesting/purposeful:* *6. About the whole group:*	*Population is not clear—what high-school students? Intent is not clear—is this to be a comparison in which an experiment is conducted with some students assigned to take Vitamin C and others not? Will we then compare the proportion who catch cold in each group, or look for an association based on observational data from surveyed students?*
What is the typical length of girls' hair?	*1. Variable of interest is clear:* *2. Population is clear:* *3. Intent is clear:* *4. Data:* *5. Interesting/purposeful:* *6. About the whole group:*	*Population of girls needs to be specified, For example, age interval of girls and where the girls are from. This question would be better if it asked "what are typical lengths of girls' hair." The phrasing "typical" suggests an average for an answer and is therefore not about the whole group.*
Is there an association between gender and ability to roll the tongue?	*1. Variable of interest is clear:* *2. Population is clear:* *3. Intent is clear:* *4. Data:* *5. Interesting/purposeful:* *6. About the whole group:*	*Population needs to be more specific.*

Part II: Write an investigative question for each of the following general topics:

1. Price of a new car

Possible answer: *What are the prices of new 2020 mid-size SUVs?*

2. Effect of listening to music and math test scores

Possible answer: *Do students at our high school who listen to music while taking an algebra test tend to score better than those students at our high school who do not listen to music while taking an algebra test?*

Exit Ticket

1. A group of biology students asked the question, "What's the fastest animal in the world?"

» Explain why this is not a well-written investigative question.

Possible answer: There is only one answer—no variability in the data.

» Rewrite the question so it would be a well-written investigative question.

Possible answer: What are typical speeds of animals found in the continental United States?

2. A coffee house owner asked the question, "How much money is spent in my coffee house?"

» Explain why this is not a well-written investigative question.

Possible answer: Population is not clear. Define "my" coffee house. Also, is the coffee owner interested in a certain time frame the coffee shop is open?

» Rewrite the question so it would be a well-written investigative question

Possible answer: How much money do customers spend between the hours of 7 a.m. and 10 a.m. at the Java House?

3. A politician asked the question, "If the election were held today, whom would you vote for?"

» Explain why this is not a well-written investigative question.

Possible answer: This is an example of a survey question to obtain data for answering an investigative question. Population is not specified.

» Rewrite the question so it would be a well-written investigative question.

Possible answer: What proportion of likely voters in Wisconsin will vote for each presidential candidate in the upcoming election?

Extension

Student Worksheet 3: Investigative Process

Ms. Brown, an administrator at a local high school of 935 students, is interested in studying the effects of playing video games on the academic achievement of students.

Formulate an investigative question that can be answered with data. Remember to clearly identify the variables of interest, the population, and the data or measurements needed. Check your investigative question against the criteria for a well-written investigative question.

Possible answer: *Is there an association between the number of hours per school day ninth grade students at the local high school play video games and their GPA at the end of the ninth grade school year?*

Create a data-collection plan including any survey questions.

Possible answer: *Randomly select 50 ninth grade students and ask each selected student how many hours per school day he/she usually plays video games. At the end of the school year, ask the principal for the GPAs of the 50 selected students.*

Describe possible data analysis questions and types of graphs and calculations needed to be performed.

Possible answer: *Construct dot plots of the amount of time playing video games and GPAs. Explore the distributions to describe the amount of time playing video games and GPAs. Construct a scatterplot of GPA vs. Video Game Hours for the 50 students. Explore the possible association of GPA and Video Game Hours by describing trend and strength (weak, moderate, strong).*

Describe how Ms. Brown might interpret the results.

Possible answer: *Does the scatterplot show a pattern/association between video game hours and GPA, or is there no pattern that emerges?*

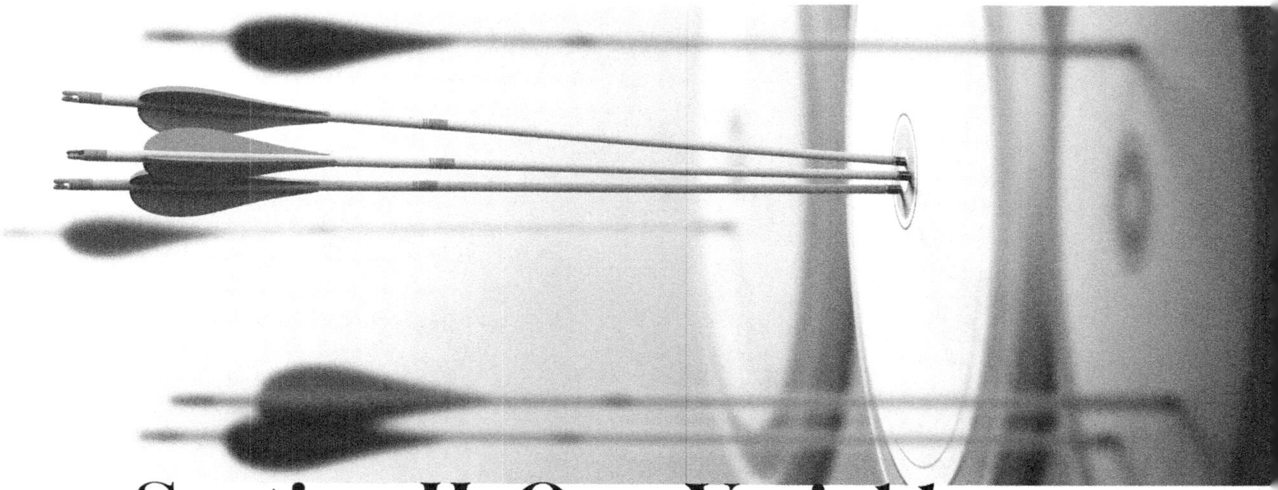

Section II: One-Variable Data Analysis

Investigation 1

Could You Be an Olympic Swimmer?

Graphical Displays

Overview

This investigation develops the concept of representing a distribution with a dot plot and a box plot and using these graphical representations to answer a statistical question. Students are asked to measure their height and arm span. They then compare the ratios of arm span to height of the class to the ratio of arm span to height of Michael Phelps. This investigation follows the four components of statistical problem solving put forth in the *Guidelines for Assessment and Instruction in Statistics Education (GAISE) Report.* The four components are: formulate a statistical question, design and implement a plan to collect data, analyze the data by measures and graphs, and interpret the results in the context of the original question. This is a GAISE Level A activity.

Instructional Plan

Brief Overview

» Have students read and discuss the scenario concerning Michael Phelps.

» Develop a statistical question concerning the distribution of arm span to height ratios.

» Discuss with students how to measure their arm span and height.

» Students collect and record their measurements and find the ratios of arm span to height.

Wandering Albatross

» Students use class data to create a box plot and a dot plot.

» Students analyze the class ratios and compare their ratios to Michael Phelps's ratio.

Hand out Student Worksheet 1.1 Height and Arm Span

Scenario

Do you know what type of bird has the largest wingspan?

The "Wandering Albatross" has been declared the *bird with the largest wingspan* among all living birds. Its wingspan, on average, is from 8.2 to 11.5 feet. Its length is from 3.51 to 4.43 feet. This wandering albatross breeds in several islands north of the Antarctic Circle and feeds off the coast of New Zealand. One albatross that was banded and followed by scientists was reported to have traveled around 3,700 miles in just 12 days.

An ornithologist is a scientist who studies every aspect of birds. One aspect an ornithologist uses to compare bird species is the ratio of a bird's wingspan to the length of its body. What is the ratio of the Albatross's wingspan to length?

Answer: *Using the maximum values for the wingspan and length, the ratio is 11.5 to 4.43, or the Albatross's wingspan is about 2.5 times its length.*

Why do you think the Albatross has such a large wingspan when compared to its length?

Possible answer: *The Albatross travels great distances and needs a greater wingspan to conserve energy and be able to glide and not flap its wings as much.*

Learning Goals

» Describe the distribution of a quantitative variable from a dot plot and box plot.

» Draw appropriate conclusions based on box plots and dot plots.

Mathematical Practices through a statistical lens

MP2. Reason abstractly and quantitatively

Statistically proficient students reason in the presence of variability and anticipate, acknowledge, account for, and allow for variability in data as it relates to a context.

Materials

Student worksheets are available at *www.statisticsteacher.org/statistics-teacher-publications/focus.*

» Measuring tapes, meter sticks, rulers, masking tape, poster paper

» Student Worksheet 1.1 Height and Arm Span

» Student Worksheet 1.2 Measuring Directions

» Student Worksheet 1.3 Sample Data

» Student Exit Ticket

» Optional: Technology to find summary statistics and construct box plots and dot plots

Estimated Time

Two 50-minute class periods—approximately one period for data collecting and a second period for the analysis of the collected data

Pre-Knowledge

Students should already be able to construct a dot plot and a box plot of a set of data.

Monarch butterfly

While an ornithologist studies birds, a lepidopterist studies butterflies and moths. The wingspan of a Monarch butterfly is from 3 ½" to 4" and its length is from 1 ¼" to 1 3/8 ".

What is the ratio of a Monarch butterfly's wingspan to length ratio?

Answer: *Using the maximum values for wingspan and length, the ratio is 4 to 1 3/8, or about 2.9. A Monarch's wingspan is almost 3 times its length.*

How does a ninth grader's arm span to height ratio compare to that of the wingspan to length ratio of the Albatross and Monarch butterfly? Do the students follow a similar pattern, in which the wingspan is many times the length?

Do you know who is the most-decorated Olympic swimmer?

Note: If time permits, show the video of Michael Phelps winning the 200m individual medley at the 2016 Rio Olympics: *www.youtube.com/watch?v=e-XGSYnhUjg*.

Michael Phelps is the most decorated Olympian of all time, with a total of 28 medals (as of the 2016 Olympics). Phelps also holds the all-time record for Olympic gold medals with 23. At the 2016 Summer Olympics in Rio de Janeiro, he won five gold medals and one silver, more than any other competitor for the fourth Olympics in a row.

Formulate a Statistical Question

Share a photo of Michael Phelps swimming that shows his arm span (one example is at *https://itraintherefore ieat.com/tag/michael-phelps*).

Discuss with your students how some people have suggested one reason Michael can swim so fast is because he has unusually long arms compared to his height. Michael stands 6'4" (193.04 cm) and has an arm span of 6'7" (200.66 cm). The ratio of his arm span to his height is 200.66/193.04 or 1.039.

Ask your students if they think his ratio is unusual.

Ask your students what might be some other reasons Michael has been so successful.

Possible answers: *Training, excellent physical shape, large hands and feet*

Ask your students if they think anyone in class has a greater ratio of arm span to height than Michael Phelps and, if so, mention that maybe they could be an Olympic champion. Encourage your students to consider the statistical question: "What is the typical arm span to height ratio of students in our class?" Also suggest they consider the question, "How does the ratio of arm span to height of Michael Phelps compare to the students in our class?"

Collect Appropriate Data

Following are three options for collecting the appropriate data.

Option 1: Students create procedures and help set up measuring stations.

Swimmer with a large arm span

Prior to placing students in groups and measuring their arm span and height, discuss some of the following points:

» Explain that they will be measuring their arm span and height and then investigating any patterns in the class *ratios* of arm span to height.

» Ask them how they could measure their arm span. Tip of finger to tip of finger? Tape a measuring tape or rulers to the wall? Make measuring strips?

» How do they plan to measure their height? Shoes off or on? Tape rulers to the wall? Or tape a measuring tape to the wall? Or their own measuring strip?

» What unit of measure should be used? How precise should the measurements be made? *Suggest the students use the metric system and measure to the nearest cm.*

After the class discussion, write the student-generated procedures on the poster paper and place near the measuring stations.

Provide each group with measuring tapes/rulers and masking tape.

Allow students time to collect their height and arm span. Draw a table on the board or set up a table on the classroom computer where students can record their arm span and height. Include Michael Phelps's measurements on the first row of the table.

Option 2: Students follow directions at teacher set-up stations.

Have an appropriate number of height measuring stations and arm span measurement stations set up before class or have students help set up the stations. You may wish to secure a tape measure to the wall, secure 2- or 3-meter sticks to the wall, or have students create their own measuring strip of paper.

Place your students into groups. Hand out Student Worksheet 1.2 Measuring Directions. This worksheet describes one method students can use to set up a measuring station for their height and arm span. If using a different method, Student Worksheet 1.2 Measuring Directions is not needed.

Discuss with your students how to set up each measuring station and how to measure their height and arm span.

Allow students time to collect their height and arm span. Draw a table on the board or set up a table/spreadsheet on a calculator/computer where students can record their arm span and height. Include Michael Phelps's measurements on the first row of the table. The students should record the ratio of their arm span to their height in the same row as their arm span and height measurements (see Table 1.1).

Option 3: If time or resources are a problem, provide students with sample data collected from the American Statistical Association Census at

Table 1.1 Sample Table

Student	Arm Span (cm)	Height (cm)
Michael Phelps	201	193
1		
2		

Table 1.2 Example

Student	Arm Span (cm)	Height (cm)	Arm Span / Height
Michael Phelps	201	193	1.04
1			
2			

School website: *ww2.amstat.org/censusatschool* (Student Worksheet 1.3 Sample Data).

Analyze the Data

After students have recorded their arm span and height, ask them to determine the ratio of the arm span to height for Michael Phelps. Then ask each student to determine their ratio of arm span to height. Record the ratios in the table (example in shown in Table 1.2) in the appropriate row.

Ask what Michael Phelps's ratio of 1.04 means.

Answer: *This means Michael Phelps's arm span is just slightly larger than his height because the ratio is just a little more than 1.*

Explain that we want to create different graphical representations of the ratios of arm span to height and use these representations to compare the class ratios to Michael Phelps's ratio of 1.04.

Visualizing the Data with Dot Plots

Refer to Student Worksheet 1.1 Height and Arm Span. Ask the students to complete questions 1 to 4 while working in their groups.

Discuss the answers.

Answers using data from Student Worksheet 1.3 Sample Data

1. What does it mean if a person's arm span to height ratio is equal to 1? Less than 1? More than 1?

Answers:

» *A ratio equal to 1 means the arm span and height of an individual are equal.*

» *A ratio less than 1 means the arm span is less than the height.*

» *A ratio greater than 1 means the arm span is greater than the height.*

2. Construct a dot plot of the class ratios of arm span to height. Include Michael Phelps's ratio.

Answer: *Figure1.1*

3. Describe the center, shape, and spread of the data.

Answer: *The center is around 0.98 (median is 0.988 and mean is 0.974), the distribution is skewed left, and the data spread from 0.829 to 1.042 with a cluster of data from about 0.96 to 1.04.*

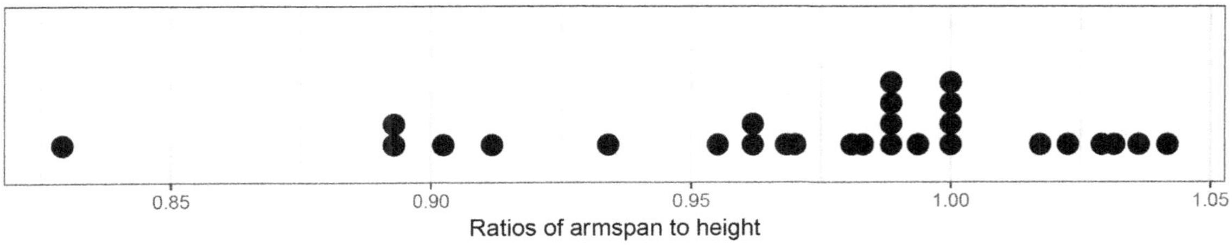

Figure 1.1: Dot plot of the class ratios of arm span to height

4. Using the dot plot, what can you conclude about the ratio of arm span to height for the students in class?

Possible answer: The shape of the distribution is skewed left with a possible outlier at 0.83. Most students have about the same height and arm span because the data center a bit less than 1.

Visualizing the Data with Box Plots

Working in their groups, ask the students to complete questions 5 and 6 on Student Worksheet 1.1.

Answers use data from Student Worksheet 1.3 Sample Data.

5. Construct a box plot of the class ratios of arm span to height. Include Michael Phelps's ratio in the class data. Use the same scale used for the dot plot.

Answers based on sample data:

» *Figure 1.2*

» *minimum = 0.829*

» *Q1 = 0.958*

» *median = 0.988*

» *Q3 = 1*

» *maximum =1.042*

6. What percent of the ratios are less than the lower quartile? What percent of the ratios are less than the upper quartile?

Answers:

» *Approximately 25% of the ratios are less than the lower quartile.*

» *Approximately 75% of the ratios are less than the upper quartile.*

Have your students answer questions 7 to 9.

Note: If students are not familiar with the concept of an outlier, demonstrate how to find an outlier using the following definition:

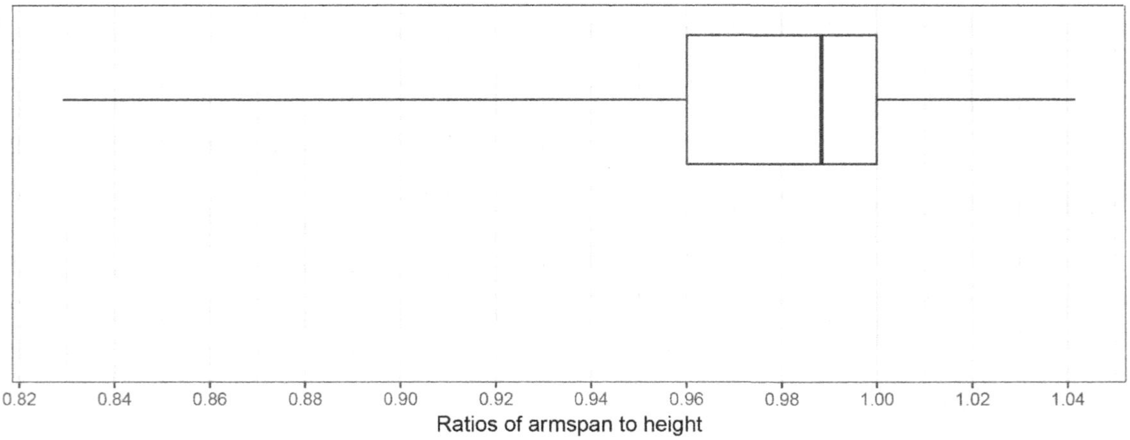

Figure 1.2: Box plot of the class ratios of arm span to height

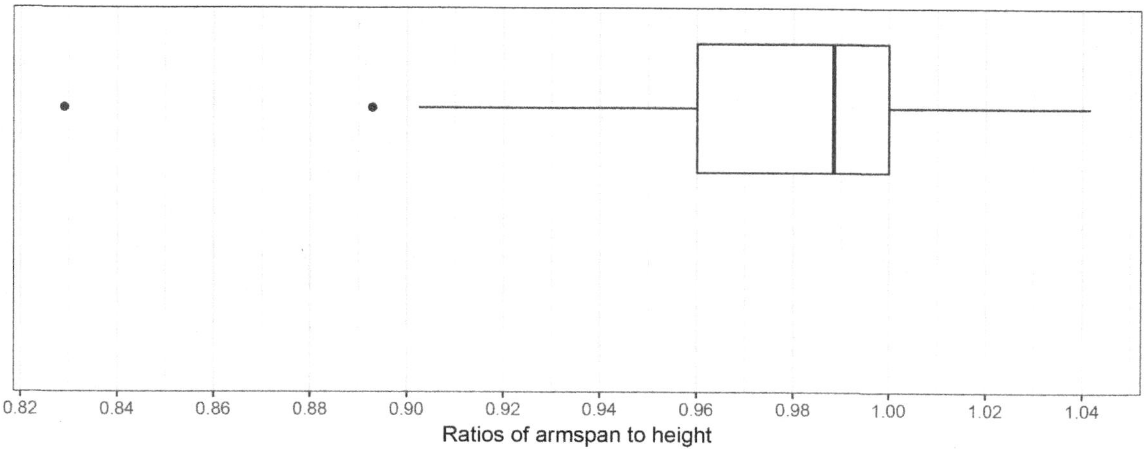

Figure 1.3: Box plot of the class ratios of arm span to height showing outliers

A data point is an outlier if it falls more than 1.5*(IQR) above the upper quartile or more than 1.5* (IQR) below the lower quartile.

Note: The interquartile range (IQR) is the difference between the upper quartile (Q3) and lower quartile (Q1).

7. Use the given definition of an outlier and determine if there are any outliers. Is Michael Phelps's ratio an outlier?

Answers:

» *There is an outlier at 0.829 and 0.893 because they are less than 0.958-1.5(1-0.958) = 0.895.*

» *Michael Phelps is not an outlier because (using the sample data) 1.042 is not greater than 1+1.5(1-0.958) = 1.063.*

8. Using the box plot, what can you conclude about the ratio of arm span to height for the students in class?

Possible answer: *Students should comment on the variability of the data, including the middle 50%. They should recognize if there are outliers and comment about how Michael Phelps's ratio compares to the class ratios. For example: "The middle 50% of the students have ratios between 0.958 and 1, telling me that about half of the class has an arm span equal to or slightly less than their height. There are two outliers, indicating these ratios are not like the rest of the class."*

9. Explain how the box plot and dot plot each helped in comparing the class data with Michael Phelps's ratio of 1.04.

Possible answer: *The box plot helps us see whether there are any outliers and the overall spread of the data. The dot plot is helpful in describing the shape and estimating the mean of the distribution.*

If there are outliers, then demonstrate how to change the box plot to show outliers. Place an asterisk or x at each outlier and then draw the segment from the quartile to the most extreme data point that is not an outlier (see Figure 1.3).

Answer using data from Student Worksheet 1.3 Sample Data (shows outliers at 0.829 and 0.893; the segment or whisker ends at 0.902).

Interpret the Results in the Context of the Original Question

Allow time for your students to complete Question 10.

Discuss the answers to Question 10. Have the students share their summary with other groups.

10. Write a summary answering the statistical question. Your summary should include how your ratio compared to others in class and to Michael Phelps's ratio. Also, how did the class ratios compare to Michael Phelps's ratio?

Possible answer: *Students should comment on the position of Michael Phelps's ratio—is it an outlier, are there students who have a ratio greater than Michael Phelps? For example, "My ratio of___ is (a lot more/less, a little more/less) compared to the rest of the class and (similar comparison statements) compared to Michael Phelps. This implies that Michael Phelps is not that unusual for his body ratio. There are probably other factors that contribute to his success. However, having arms much shorter than one's height might not make the best swimmer."*

Exit Ticket

How much of a tip should you leave at a restaurant? Currently, the standard tip is between 15% and 20% of the pre-tax amount of the bill. Sara, a high-school senior was working part- time at a local diner. Most items on the menu were from \$6.50 to \$12.00, and most of the tables she served had either one or two diners. One Saturday, Sara kept track of the tips she earned for 25 tables she served. Below (Figure 1.4) is a box plot and dot plot of the tip amounts.

Using the two graphical representations, describe the distribution of the amount of the tips Sara received on the Saturday. Include in your description the center and spread of the data.

Possible answer: *The median tip value was \$1.70, and the mean was \$2.08. The distribution is skewed right, with two outliers at \$5.75 and \$6.00. Most of Sara's tips were between \$1.00 and \$2.50.*

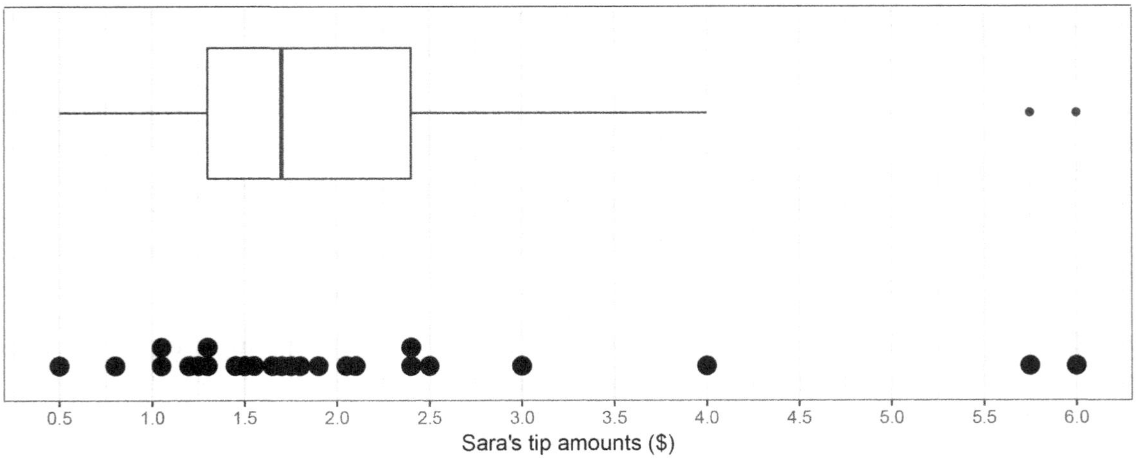

Figure 1.4: Box plot and dot plot of the tips Sara earned for 25 tables she served

Further Exploration

- Have students measure the distance from the top of their head to their chin. Have them investigate the relationship of the ratio of this distance to height for the students in their class. For an average adult, the total height is equivalent to 7 to 7.5 heads tall. How do the students' ratios compare to this standard?
- Have students measure the distance from the bottom of their nose to the outside corner of their right eye and the length of their right ear. For an average adult, the ratio of the distance to the eye and length of the ear is about 1. How do the students' ratios compare to this standard?
- Have students measure the width of their head and length of an eye. For an average adult, the ratio of head width to eye length is between 4 and 5. How do the students' ratios compare to this standard?

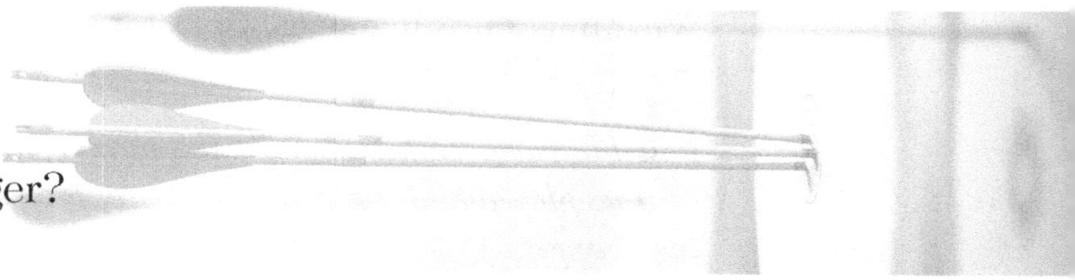

Investigation 2

Are Baseball Games Taking Longer?

Comparing Multiple Groups

Overview

This investigation asks students to use graphical displays—box plots and dot plots—to describe the distribution of a quantitative variable (length of major league baseball games). They compare the distributions of the length of games in baseball seasons from three decades using parallel box plots and summary statistics. The investigation follows the four components of statistical problem solving put forth in the *Guidelines for Assessment and Instruction in Statistics Education (GAISE) Report*. The four components are formulate a statistical question, design and implement a plan to collect data, analyze the data by measures and graphs, and interpret the results in the context of the original question. This is a GAISE Level A/B activity.

This investigation is based on an article in *USA Today*, "Average MLB Game Time Rises to Record High," published October 2, 2017. Read it at *www.usatoday.com/story/sports/mlb/2017/10/02/average-mlb-game-time-record-3-hours-5-minutes-this-season/106225166.*

Instructional Plan

Brief Overview

» Have students read and discuss the scenario about the length of baseball games from three years.

» Formulate a statistical question comparing the distribution of the length of baseball games in three years.

» Have students construct a box plot and dot plot for each of the three years.

» Construct parallel box plots and dot plots. Use the plots to compare the length of games in the three years.

Scenario

Have you been to a major league baseball (MLB) game recently? Or maybe watched one on television? What did you think about the length of the game? Was it too long, too short, or just about right?

The biggest complaint of many baseball fans—both young and old—is that the pace of the game is too slow. Throughout the decades, the length of major league games seems to have increased. Fans give all sorts of reasons for why the games might be getting longer. Some say it's due to more and longer TV commercial breaks; others say it's because of the multiple mid-inning pitching changes; still others suggest it's due to the use of replay by umpires to decide close calls.

What might be some other reasons the length of MLB games would increase?

Possible answer: *Constant stepping out of hitters from the batters' box and pitchers taking an inordinate amount of time between pitches with no one on base.*

What might be some suggestions for speeding up MLB games?

Possible suggestions: *Shorten the number of innings played and limit the number of pitching changes.*

Formulate a Statistical Question

Discuss the following with your students:

For a statistical study, a small group of high-school students wanted to investigate how the length of major league baseball games has changed over time. They decided to look

Learning Goals

- » Describe the distribution of a quantitative variable from a dot plot and box plot.
- » Compare the distributions of a quantitative variable using parallel box plots and summary statistics.

Mathematical Practices Through a Statistical Lens

MP6. Attend to Precision

Statistically proficient students are precise about choosing the appropriate analyses and representations that account for the variability in the data. They display carefully constructed graphs with clear labeling. As students interpret the analysis of the data, they are precise with their terminology and statistical language.

Materials

Student worksheets are available at *www.statisticsteacher.org/statistics-teacher-publications/focus.*

- » Student Worksheet 2.1 Length of Baseball Games
- » Student Worksheet 2.2 Analyzing the Times
- » Student Exit Ticket
- » Optional: Worksheet 2.3 to be used with Addition Ideas Section: Directions to Use New Zealand Census at School Website, *http://new.censusatschool.org.nz/tools/random-sampler*
- » Optional: Technology to find summary statistics and construct dot plots and box plots

Estimated Time

Two 50-minute class periods

Pre-Knowledge

Students should be able to describe the shape, center, and variability of a distribution of a quantitative variable. They should also know how to construct a dot plot and box plot of a quantitative variable.

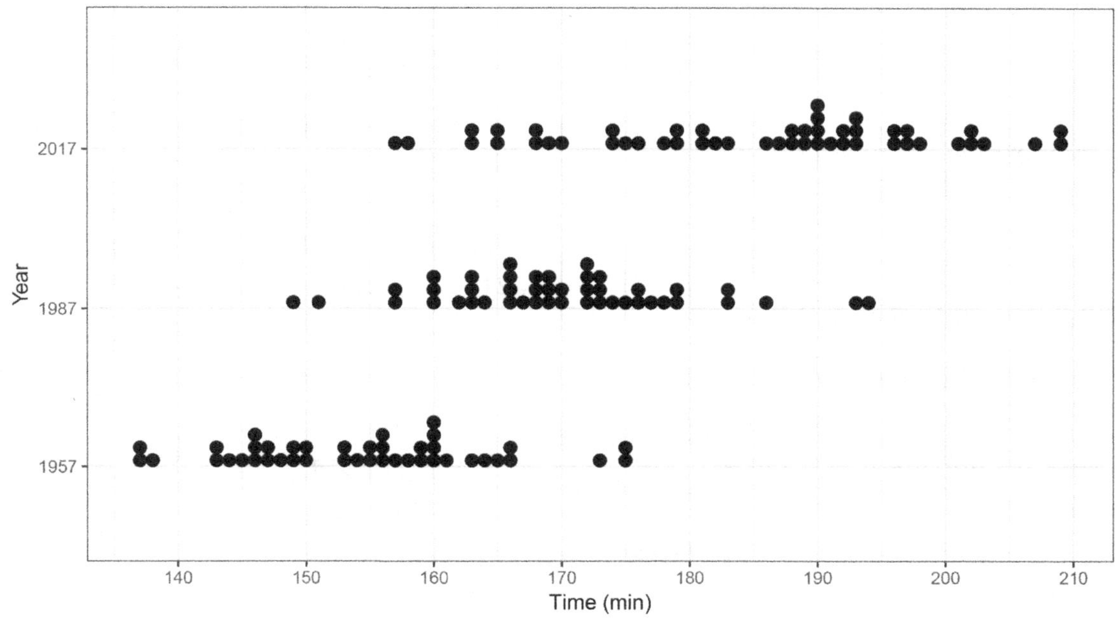

Figure 2.1: Dot plots for each of the three years included in the random sample

at one year toward the end of each of three decades—1950s, 1980s, and 2010s. Rather than try to manage the analysis of all the games played in 1957, 1987, and 2017, they decided to take a random sample of the length of the games from those three years.

The students used the data to investigate the statistical question, "Has there been an increase in the length of time to play regular season major league baseball games in 1957, 1987, and 2017?"

Collect Appropriate Data

Distribute Student Worksheet 2.1 Length of Baseball Games.

Point out that the data collected are from 50 randomly selected games from 1957, 46 games from 1987, and 43 games from 2017. The data were collected from samples of 9-inning games. Any games that lasted more than 9 innings, had rain delays, or were shortened due to weather were not included.

The data can be found at *www.baseball-reference.com/leagues/MLB/misc.shtml.*

Analyze the Data

Hand out Student Worksheet 2.2 Analyzing the Times.

Place students into groups of four and have the groups complete questions 1 to 7 on Student Worksheet 2.2 Analyzing the Times.

1. Work with members of your group to construct a dot plot of the sample lengths for baseball games in 1957. Use a scale from 130 minutes to 210 minutes.

Possible answer: *Figure 2.1 shows the dot plots for each of the three years.*

2. Using the dot plot for the year 1957, estimate the center of the distribution. Describe the spread of the data.

Possible answer: *The center of the distribution is approximately 155 min. The spread of the* *distribution goes from approximately 135 to 175 min. The distribution mounds up around 160 min.*

3. Work with members of your group to construct a dot plot of the sample lengths for baseball games in 1987. Place the dot plot above the dot plot for year 1957.

Possible answer: *See Figure 2.1.*

4. Using the dot plot for the year 1987, estimate the center of the distribution. Describe the spread of the data.

Possible answer: *The center of the distribution is approximately 170 min. The spread of the distribution goes from approximately 145 to 195 min. The distribution mounds up around 170 min.*

5. Work with members of your group to construct a dot plot of the sample lengths for baseball games in 2017. Place the dot plot above the dot plots for the years 1957 and 1987.

Possible answer: *See Figure 2.1.*

6. Using the dot plot for the year 2017, estimate the center of the distribution. Describe the spread of the data.

Possible answer: *The center of the distribution is approximately 185 min. The spread of the distribution goes from approximately 155 to 210 min. The distribution mounds up around 190 min.*

7. Using the three dot plots, what observations can you make concerning the length of the games in the three years? Comment on the center and spread of each dot plot.

Possible answer: *The centers have increased from 1957 to 2017. The minimums and maximums have also increased from 1957 to 2017.*

Have the students answer questions 8 to 10 and then discuss their answers.

8. To help compare the length of the games, work with members of your group to construct a box plot for the sample lengths of baseball games in each of the three years 1957, 1987, and 2017. Place the three box plots on the same number line with a scale from 130 minutes to 210 minutes to form parallel box plots.

Answer: *Figure 2.2 shows the box plots for each of the length of the games in the years 1957, 1987, and 2017.*

Using the parallel box plots of the samples of lengths of games, how do the length of major league games in 1957, 1987, and 2017 compare? Comment on the center and spread for each distribution.

Possible answer: *The length of the games appears to have been increasing from 1957 to 2017. Seventy-five percent of the lengths in 2017 are greater than all the games in the 1957 sample and greater than 75% of the 1987 games. The median of the 2017 games is greater than both the 1987 games and 1957 games. At least 75% of the 1987 games are greater than 75% of the 1957 games.*

9. What advantages and disadvantages do box plots have over dot plots for making comparisons between multiple groups of data?

Possible answer: *The parallel box plots make it easy to compare the centers (medians) and overall spread of the data. The dot plots give the overall shape of each distribution, which can be helpful when comparing distributions.*

Interpret the Results in the Context of the Original Question

Have the students answer Question 10.

10. Based on the three dot plots and parallel box plots you constructed, do

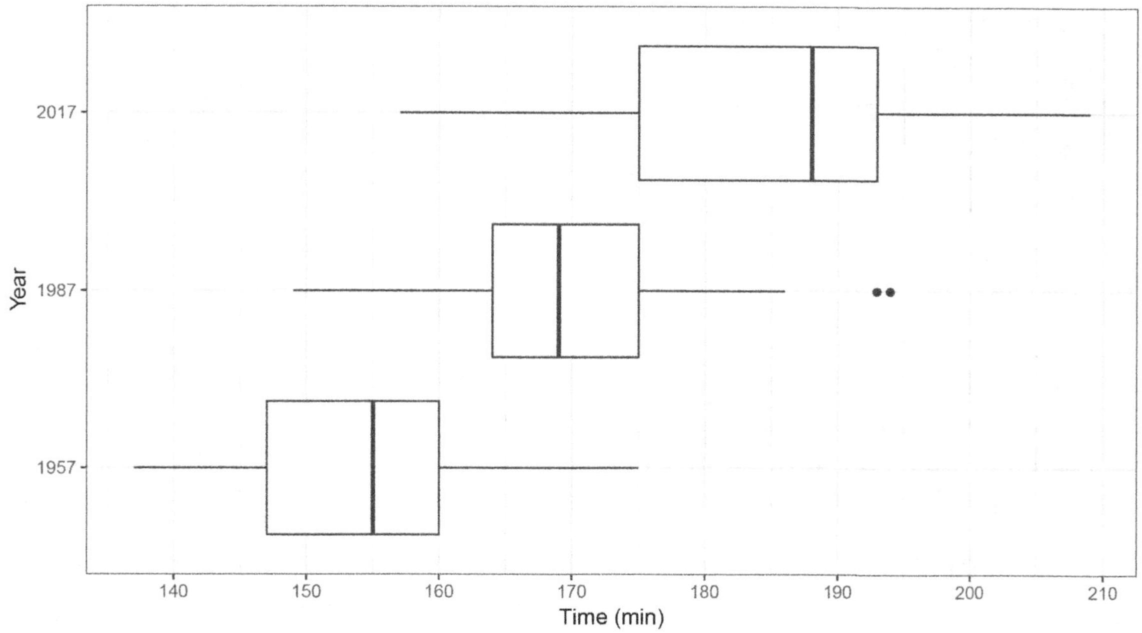

Figure 2.2: Box plots for each of the three years included in the random sample

you think the length of the games has changed by any meaningful amount? Explain your thinking.

Possible answer: *Yes, the median for 2017 is 17 minutes longer than the 1987 games and more than 30 minutes longer than the 1957 games. All the 2017 games are longer than the bottom half of the 1957 games. Seventy-five percent of the 2017 games are longer than the 1957 games.*

Additional Ideas

New Zealand Census at School hosts the random sampler (*http://new.censusatschool.org.nz/tools/random-sampler*) for international, New Zealand, and US data that have been "cleaned"—data entered incorrectly have been removed.

Using the random sampler at the New Zealand Census at School website, take a random sample of at least 75 ninth graders. Take the responses from Question 8—"What is the main method of transportation you typically use to get to school?"—and the answer to Question 9—"How long does it usually take you to travel to school?"—and create a table showing a summary of the responses. Then, using the table, create graphical representations that will help answer the statistical question, "How do the length of times to get to school compare for ninth graders who walk, ride a bus, or ride in a car?"

Directions for using New Zealand Census at School are on Student Worksheet 2.3.

Exit Ticket

On February 28, 1983, the final episode of M*A*S*H* aired on CBS. As of 2017, it remained the most-watched TV series finale. It was estimated that at least 105.9 million people watched this last show with a household rating of 60.2%. A 60.2% household rating means 60.2% of all households—homes with a TV set—were tuned to the final episode of M*A*S*H. Source: *https://en.wikipedia.org/wiki/List_of_most_watched_television_broadcasts_in_the_United_States*

The parallel box plots in Figure 2.3 show the household ratings in the past four decades for most of the top 66 TV series finale broadcasts. The data were sorted by the three major networks—ABC, CBS, and NBC—and a box plot was constructed of the household ratings for finale TV shows for each of the networks. Data are from 2017.

1. Describe the distribution of household ratings for each of the three networks. Include the center and spread of the data in your description.

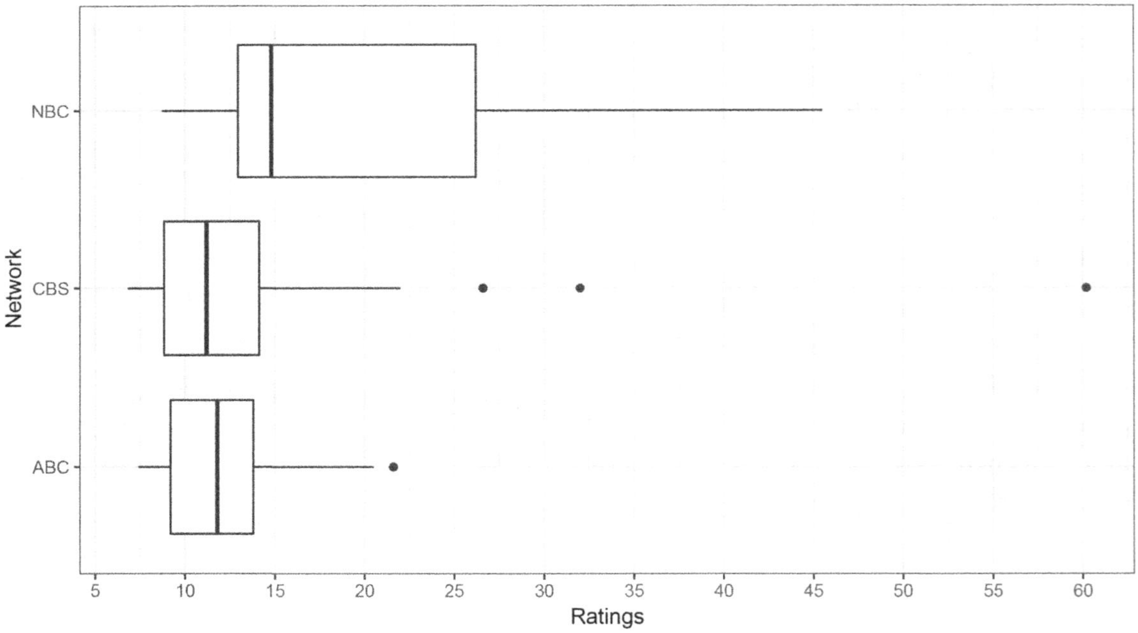

Figure 2.3: Box plots showing the household ratings in the past four decades for most of the top 66 TV series finale broadcasts

Possible answer:

- *ABC: The center has a rating of about 12 with an outlier at approximately 22. The spread of the data is small with the middle 50% of the data between 8 and 14, or an IQR equal to 6.*
- *CBS: The center has a rating of around 11 with three outliers at approximately 27, 33, and 61. The spread around the median is small with an IQR of approximately 6.*
- *NBC: The center has a rating around 14 with no outliers. NBC has a larger spread than the other two networks with an IQR of approximately 15.*

2. Which network would you rank as the top network when comparing the household ratings for the top 66 TV series finale broadcasts? Give reasons for your answer.

Possible answer: *NBC ranked highest because it has the highest median and the top 50% of their show ratings were greater than 75% of ABC's and CBS's shows.*

Further Explorations and Extensions

Common Core State Standard 7.SP.A.3 states: "Informally assess the degree of visual overlap of two numerical data distributions with similar variabilities, measuring the difference between the centers by expressing it as a multiple of a measure of variability."

Informally investigate this standard looking at the difference between the medians in parallel box plots and informally deciding if the medians represent a "significant" difference.

One informal method that can be used to determine if the difference between the medians represent a "significant" difference is to assess the degree of visual overlap of the box plots. This informal assessment can be done by determining how many IQRs separate the medians.

For example, the median of 1987 and the median of 1957 differ by 14 minutes. The IQR is 12 minutes for 1987 and 13 minutes for 1957. Divide the difference between the medians by the higher of the two IQRs, 14/13 or 1.08 IQR's. This doesn't appear to be an extremely large difference.

Have students compare the difference of the medians for 2017 with 1987 and 1957 to see if there is a meaningful difference between the medians.

Investigation 3

How Good Is Your Memory?

Standard Deviation

Overview

This investigation builds on the description of the spread of a distribution by developing a key measure of variability—the standard deviation. A procedure to calculate standard deviation is developed using data from the results of a memory test. Students use the mean and standard deviation to compare fourth grade memory test completion times (from the Census at School website) to their class memory test results.

This investigation follows the four components of statistical problem solving put forth in the *Guidelines for Assessment and Instruction in Statistics Education (GAISE) Report*. The four components are formulate a statistical question, design and implement a plan to collect data, analyze the data by measures and graphs, and interpret the results in the context of the original question. This is a GAISE Level B activity.

This investigation is based on lessons from *Exploring Measurements* by Peter Barbella, James Kepner, and Richard Scheaffer, published in 1994 by Dale Seymour Publications as part of the Quantitative Literacy Series. Though out of print, the book listed is available through book resale sites and on Amazon.com.

The data used in this investigation were collected from the Census at School website using the random sampler at *ww2.amstat.org/CensusAtSchool.*

Instructional Plan

Brief Overview

» Formulate a statistical question concerning comparing times to complete a memory test of your class and a group of fourth graders from around the United States.

» Have students take the memory test on the Census at School website.

» Develop a procedure to calculate the standard deviation.

» Use the mean and standard deviation to compare the distributions of times to complete the memory test.

Scenario

Do you remember playing the Memory game or the game Concentration when you were in elementary school? This game usually consisted of a deck of pairs of matching cards. The cards were spread out on a table face down in rows. Players took turns turning over a pair of cards. If the cards matched, then the player kept the pair. If the cards did not match, they were returned face down to their positions, and it was the next player's turn. After all the cards had been turned over and the pairs found, the player with the most pairs won the game.

How do you think you would do playing this game now? If you were playing the game by yourself, do you think you could find all the matching pairs in a very short period of time?

Learning Goals

- Develop the measure of variability and standard deviation and interpret the standard deviation in context.
- Compare two or more distributions using the mean of the distributions as a measure of center and the standard deviation of the distributions as a measure of spread.

Mathematical Practices Through a Statistical Lens

MP7. Look for and make use of structure

Statistically proficient students look closely to discover a structure of pattern in a set of data as they attempt to answer a statistical question.

Materials

Student worksheets are available at *www.statisticsteacher.org/statistics-teacher-publications/focus.*

- Computers or tablets with internet access
- Software or statistical apps to find summary statistics and construct dot plots
- Student Worksheet 3.1 Directions for Taking the Memory Test at the Census at School Website
- Student Worksheet 3.2 Fourth Grade Memory Test Times
- Student Worksheet 3.3 Memory Test Investigation
- Optional: Student Worksheet 3.4 Sample Ninth Grade Memory Test Times (use if not having students take the memory test)
- Student Worksheet 3.5 Standard Deviation Match
- Student Exit Ticket

Estimated Time

Two 50-minute class periods

Pre-Knowledge

Students should be able to estimate the mean of a distribution visually and calculate and interpret the mean of a distribution.

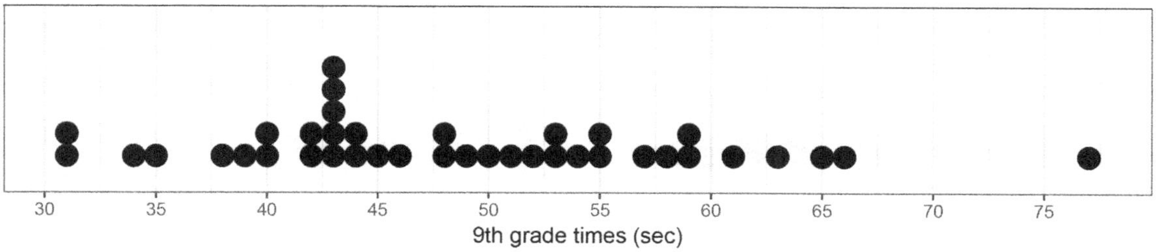

Figure 3.1: Ninth grade memory test completion times (sec.)

Formulate a Statistical Question

Explain to students that they will be taking a memory test on the Census at School website. The times to complete the test will be recorded and compared with fourth graders' times from around the United States. Do the students in class think they can beat the fourth grade completion times? Ask your students to consider the statistical question, "How do the memory test completion times of our class compare to the memory test completion times for fourth graders from around the United States?"

Collect Appropriate Data

Note: If your students are not taking the memory test, then have them use the ninth grade data on Student Worksheet 3.4. The data were generated using the random sampler at the Census at School website. The data will be used to provide examples and answers to the questions.

Distribute Student Worksheet 3.1 Directions for Taking the Memory Test at the Census at School Website.

Demonstrate how to take the memory test on the Census at School website:

Go to the Census at School website at *ww2.amstat.org/CensusAtSchool.*

Choose the Student Section.

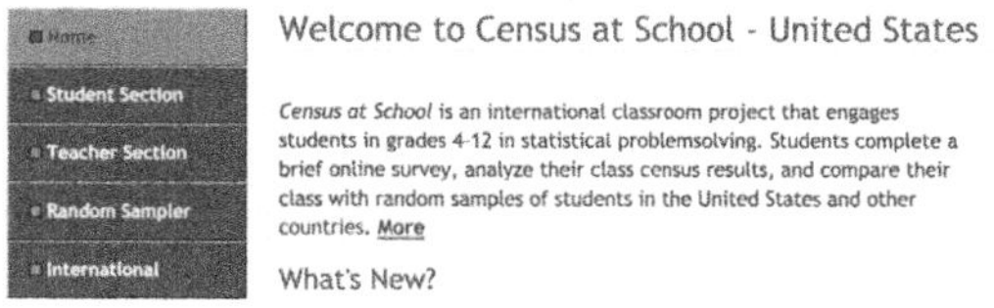

Choose the memory test on the right of the screen.

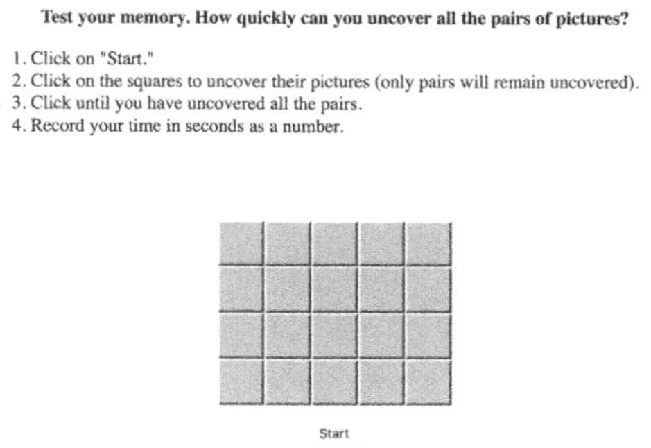

Press start and begin selecting boxes. Continue until all the matches have been found.

Direct your students to take the memory test. When they have completed the test, have them record their times as reported at the end of the test on the class dot plot and in the classroom calculator / spreadsheet / statistical software.

Note: While students are taking the memory test, draw a number line on the board for the class dot plot.

Ask your students to post their memory test results on the class dot plot.

Note: If your students did not take the memory test, then display the dot plot in Figure 3.1.

Answer using data from Worksheet 3.4 9th Grade Results.

Ask students to estimate the center of the distribution of their memory test times.

Possible answer: *Using the sample data on Worksheet 3.4, the center is approximately 50 seconds.*

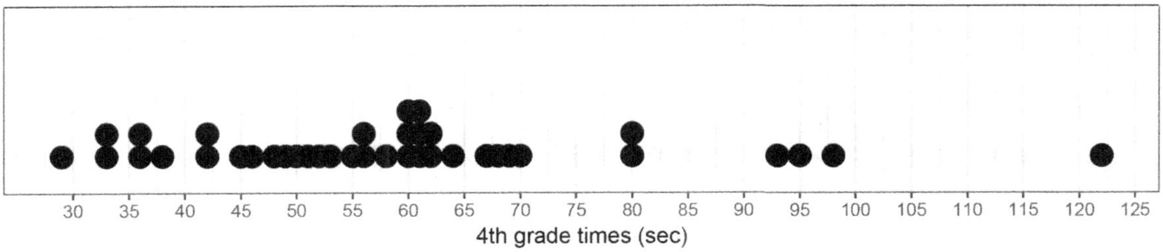

Figure 3.2: Fourth grade memory test completion times (sec.)

Analyze the Data

Distribute Student Worksheet 3.2 Fourth-Grade Memory Test Times.

Distribute Student Worksheet 3.3 Memory Test Investigation.

Ask your students to complete problems 1 to 3 on Worksheet 3.3 Memory Test Investigation.

1. To help answer the statistical question—"How do the memory test completion times of our class compare to the memory test completion times for fourth graders from around the United States?"—construct a dot plot of the fourth grade memory test times.

***Sample answer:** Figure 3.2.*

2. Describe the distribution of the fourth grade memory test times. Your description should include an estimate of the distribution's center and a description of the spread around the center.

***Possible answer:** The distribution centers around 60 sec. and has a spread from about 25 sec. to 120 sec., with much of the data from 35 sec. to 70 sec.*

3. Copy the dot plot of the class memory completion times above the dot plot of the fourth grade completion times.

***Sample answer:** Figure 3.3*

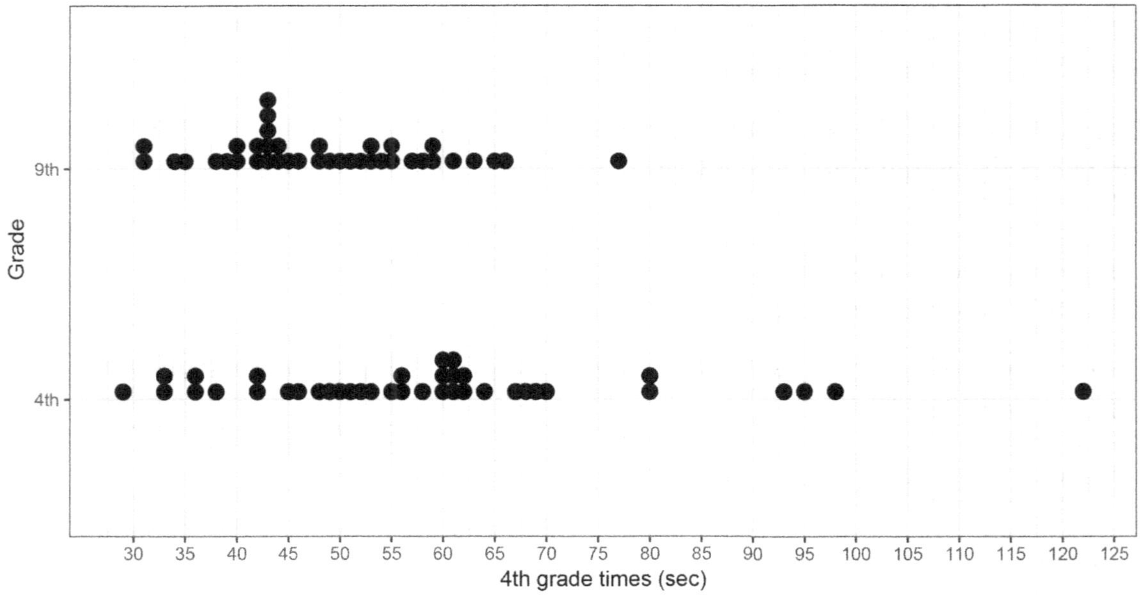

Figure 3.3: Fourth grade memory test completion times (sec.) with ninth grade memory test completion times (sec.) above.

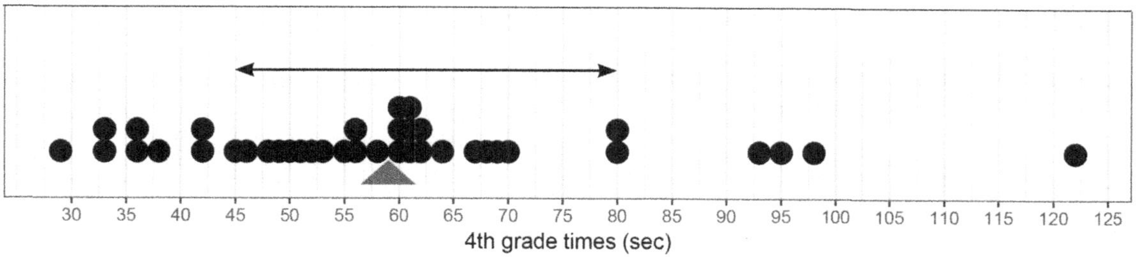

Figure 3.4: Fourth grade memory test times with the mean marked with a Δ

4. Using the dot plot of class memory completion times and the fourth grade dot plot of completion times, how do the two distributions compare?

Possible answer: *The center of the class times distribution is lower than the center of the fourth grade times. The distribution of class times is less spread out than the distribution of the fourth grade times. It appears the class times were typically better than the fourth grade times.*

When describing a distribution and comparing distributions, it is often useful to give a specific center such as the median or mean and a value that describes the variation around that center. The interquartile range (IQR) is one measure of variability that describes the variability around the median (see Investigation 2).

The next set of questions introduces a measure of variability that describes the variation about the mean. Have students work in their groups to complete questions 5 to 11.

5. Use technology (graphing calculator, spreadsheet, or app) and enter the fourth grade memory test times into a list. Use the technology to find the mean of the fourth grade memory test times. Mark the mean on the dot plot with a Δ, indicating the mean or the balance point of the distribution.

Answer: *Figure 3.4*

6. Draw an arrow from the mean of 59 sec. to the data point 45 sec. This arrow shows the distance the point 45 is below the mean. This distance is called the *deviation* from the mean.

Answer: *See Figure 3.4.*

7. Find the deviation from the mean for this data point (45) by subtracting the mean from the value of the data point.

Answer: *45-59 = -14*

8. This deviation is negative. What does this tell you about the data point in relation to the mean?

Answer: *45 sec. is 14 sec. below the mean of 59 sec.*

9. Draw an arrow from the mean of 59 sec. to the data point 80 sec. Find the deviation for the data point 80. This deviation is positive. What does this tell you about the data point in relation to the mean?

Answer: *80-59= 21 sec. 80 sec. is 21 sec. above the mean of 59 sec.*

10. Use technology (list on a graphing calculator or spreadsheet) to find the deviation from the mean for all the fourth grade times. If using a spreadsheet or lists in a graphing calculator, set a formula for a column that takes all the values of the fourth grade times minus the mean of 59.

Store these deviations in another list or column.

Table 3.1

4th-Grade Times (sec.)	Deviations from the Mean
68	9
80	21
49	-10
55	-4
69	10
62	3
98	39
42	-17
64	5
33	-26
58	-1
Etc.	Etc.

Table 3.2

4th-Grade Times (sec.)	Deviations	Squared Deviations
68	9	81
80	21	441
49	-10	100
55	-4	16
69	10	100
62	3	9
98	39	1521
42	-17	289
64	5	25
33	-26	676
58	-1	1
Etc.	Etc.	Etc.

Sample of some of the answers (see Table 3.1):

11. Use technology to find the sum of all the deviations. Why does this value make sense?

Answer: *0, the mean is the balance point of the distribution, so the sum of the negative and positive deviations equals zero.*

Explain to students that the objective is to find a number that summarizes all the deviations. Often, the mean is used as a summary of the data. Since the sum of the deviations is zero, the mean of the deviations would be zero. The mean of zero does not summarize the spread of all deviations from the mean. So, we need to deal with the negative deviations. We could take the absolute value of all the deviations or square of all the deviations—both turn all the negative deviations positive. If we find the absolute value of all the deviations and then the mean of these absolute deviations, we have found the value *MAD* (Mean Absolute Deviation). If we square the deviations instead of finding the absolute value, then the mean of these squared deviations is the *variance*. The *standard deviation* is the square root of the variance.

Note: For more information about MAD, see the Further Explorations and Extensions at the end of this investigation. Also, see the ASA's *Bridging the Gap* Investigation 3.4.

The population *standard deviation* is represented by the Greek letter sigma σ and it is a measure of variability or spread of data around the mean μ of a population.

Ask your students to complete questions 12 to 15.

To find the population standard deviation—a measure that summarizes the spread of all the deviations or the typical deviation from the mean, complete the following steps using technology and the list or column of the fourth grade completion times.

12. Use technology and square each deviation found in Question 10. Store the squares in another column or list.

Sample answer: *Table 3.2*

13. Use technology and find the sum of the squared deviations.

Answer: *sum = 14428 sec.*^{2}

14. Use technology and find the mean of the squared deviations.

Answer: 369.95 sec.^{2}

15. Take the square root of the mean of the squared deviations.

Answer: 19.23 seconds

Explain that this value is called the *population standard deviation*. It is a measure used to describe the amount of variation or spread of a set of data values around the *population mean*. The fourth grade completion times had a mean of 59 seconds with a standard deviation of 19.2 seconds. This can be interpreted as the typical distance the data points are from the mean is approximately 19.2 seconds.

Point out to students that graphing calculators, spreadsheets, computer apps, and statistical software have a built-in standard deviation function. If the built-in function reports two standard deviation values, one is the symbol σ (sigma) for the population (division of the sum of the squared deviations by n) and the other is the letter s for the sample (division of the sum of the squared deviations by n-1). If we are using data collected from a sample, then the value of s should be used in describing a distribution. The symbol s represents the sample standard deviation.

Note: Careful study has shown using n-1 when calculating the standard deviation for a sample of data gives the best estimate for the standard deviation of the population.

See Further Explanations and Extensions at the end of this investigation for the standard deviation formulas.

Ask your students to complete questions 16 to 19.

16. Enter the class times into a calculator, spreadsheet, or statistical software. Use the built-in standard deviation function and find the mean and standard deviation of your class completion times. Since the class is taken to be a population in this investigation, report the population standard deviation.

Answer based on given ninth grade times: *mean = 47.9 sec. and standard deviation = 9.86 sec.*

17. Interpret the mean and standard deviation of your class memory test completion times.

Answer based on given ninth grade times: *The mean of 47.9 is the balance point of the distribution. If all the ninth graders had the same completion time, it would be 47.9. The standard deviation of approximately 9.9 seconds is the typical distance the data points are from the mean of 47.9 seconds.*

18. Compare the mean of your class completion times to the mean of the fourth grade completion times.

Answer: The mean of the fourth grade times is 59 sec., and the mean of the ninth grade completion times is 47.9 sec. The ninth grade completion times are faster on average than the fourth grade completion times.

19. Interpret and compare the standard deviation of your class times to the standard deviation of the fourth grade times. What does the value of the smaller standard deviation indicate?

Answers will vary, but based on given ninth grade sample times: *The standard deviation for the ninth grade times is much smaller than the standard deviation for the fourth grade times. The distribution of ninth grade times is not as spread out around its mean as the fourth grade times are around its mean.*

Interpret the Results in the Context of the Original Question

Ask your students to complete questions 20 and 21.

20. Using the results of your study of the fourth grade completion times and your class completion times, write a summary of your answer to the statistical question: "How do the memory test completion times of our class compare to the memory test completion times for fourth graders from around the United States?"

Answer based on given ninth grade times: *The mean of the ninth grade group is much lower than the fourth grade group—47.9 seconds compared to 59 seconds. The standard deviation of the ninth grade times is 9.9 seconds, while the fourth grade standard deviation is 19.2 seconds. This means the spread of the data in the ninth grade distribution is much less than the fourth grade, which means the times in the ninth grade are not as spread out as the fourth grade times. Based on this summary, the ninth grade class is generally able to complete the memory test in less time than the fourth grade group and, as a class, more consistently.*

21. The fourth grade times had an outlier at 122 seconds. Delete this point from the list/column of the fourth grade times and recalculate the mean and standard deviation. What effect did the outlier have on the mean and standard deviation?

Answer: *Both the mean and standard deviation decrease; the mean is now 57.3 sec. and the standard deviation is 16.3 sec.*

Summary

Ask your students what standard deviation measures.

Possible answer: *Standard deviation measures the spread of a data distribution. It measures the typical distance between each data point and the mean.*

Handout Student Worksheet 3.5 Matching Standard Deviation to Dot Plots.

Explain that the worksheet is designed for students to match each dot plot with the appropriate distribution number. Encourage them to estimate, rather than use the formula.

Additional Ideas

Have the class answer at least one of the questions 26, 27, 31, or 32 on the questionnaire from the Census at School website. (See the Teacher Resources section of this publication for more information about the Census at School website). Use the Random Sampler and select a random sample of at least 40 fourth graders (or another grade of your choice) who answered at least one of the questions 26, 27, 31, or 32. Compare the class results to the sample of 4th graders' results.

Note: Students should use *s*, the sample standard deviation in their analysis.

Census at School Questions:

26. How many hours of sleep per night do you usually get when you have school the next day?

27. How many hours of sleep per night do you usually get when you don't have school the next day?

31. About how many text messages did you send yesterday?

32. About how many text messages did you receive yesterday?

Student Worksheet 3.5 Matching Standard Deviation to Dot Plots

Consider the following group of dot plots and summary statistics. Each of the summary statistics for each data set 1 to 5 corresponds to one of the dot plots, lettered A to E.

In the "Dot Plot" row of the table, write the letter of the matching dot plot next to the appropriate summary statistics and explain how you made your choice.

Summary	1	2	3	4	5
Mean	77.6	71	81.4	53.7	54.6
Median	80	80	80	50	50
Standard Deviation	9.3	19.8	8.5	21.2	9.7
Dot Plot					

A

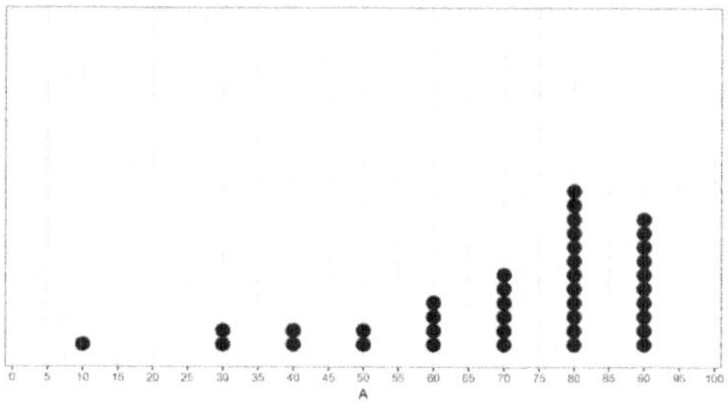

D

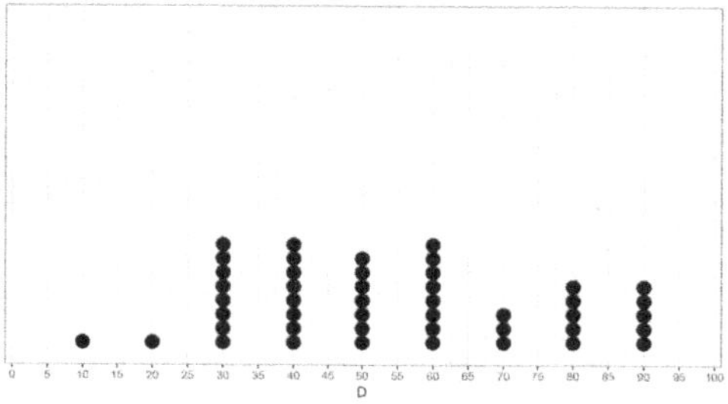

B

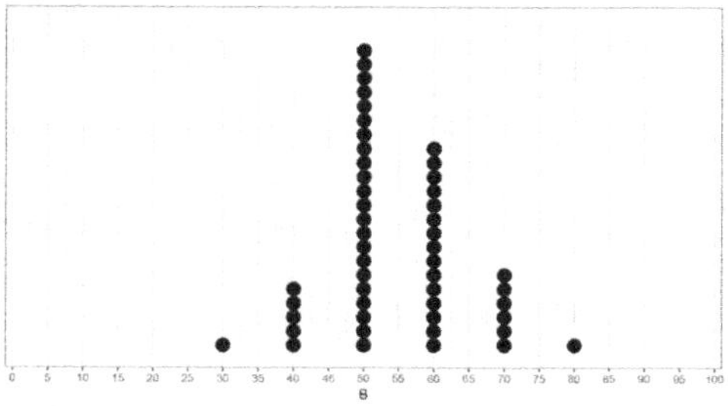

E

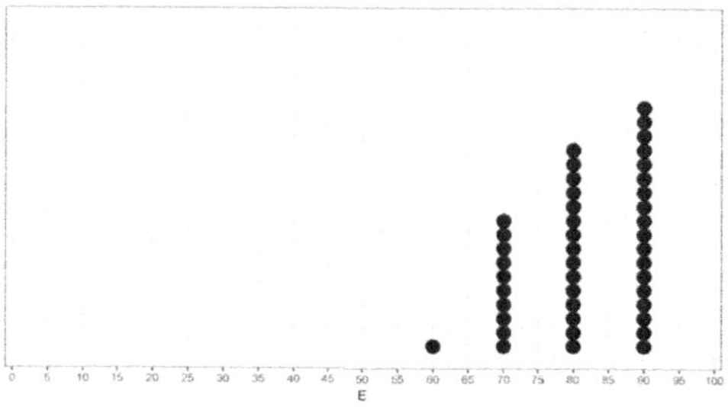

C

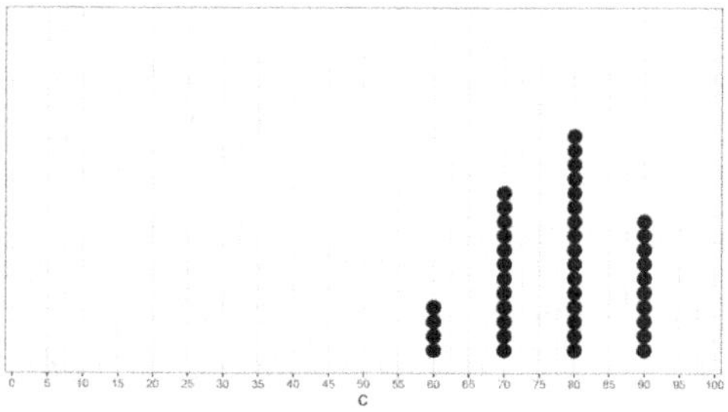

Answer: Table 3.3

Table 3.3

Summary	1	2	3	4	5
Mean	77.6	71	81.4	53.7	54.6
Median	80	80	80	50	50
Standard Deviation	9.3	9.7	8.5	21.2	9.7
Dot Plot	C	A	E	D	B

Exit Ticket

An algebra teacher wanted to determine whether ninth grade algebra students scored better when they took a math test in silence or when Mozart was being played. She randomly divided the students into two groups. One group took an algebra test in silence and the other group took the same test while a Mozart symphony was quietly playing in the room. The mean and standard deviation for the Mozart group were 55% and 17.4%, respectively, and the mean and standard deviation for the Silence group were 50.5% and 16.5%, respectively.
Data from Core Math Tools: www.nctm.org/coremathtools

The distribution of test scores from both groups is shown below.

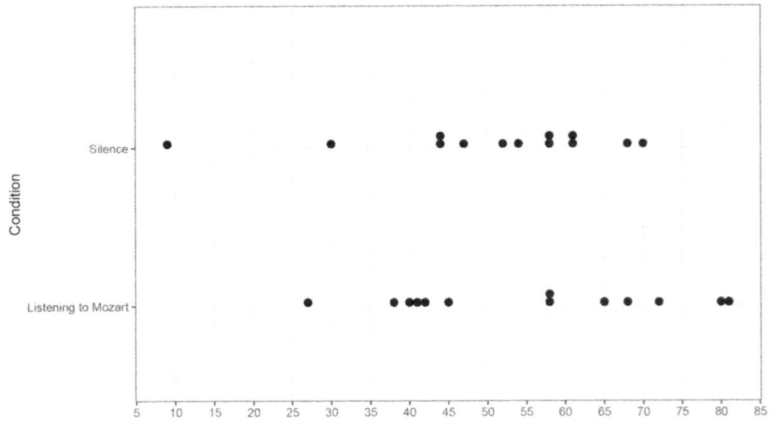

Figure 3.10: Distribution of test scores

1. Interpret the mean and standard deviation of the group that listened to Mozart.

Answer: *The mean of 55% is the balance point of the distribution. The sum of the deviations from 55% will be zero. The standard deviation of 17.4% is the typical distance a score on the math test is from the mean of 55%.*

2. Using the distribution and estimates for mean and standard deviation, did the ninth grade students in the Mozart group perform better on the math test than the group of ninth graders in the silence group?

Answer: *The scores for the ninth graders who listened to Mozart are slightly better than the scores for the group that took the test in silence. The mean of the Mozart group is around 55%, while the mean of the silence group is approximately 50%. The variation of both distributions is about the same. The Mozart group had the three highest scores, and the silence group had the lowest score.*

Further Explorations and Extensions

1. Present the formula for the population standard deviation: $\sigma = \sqrt{\frac{\sum (x-\mu)^2}{N}}$
where

x is a value in the population

x-μ is a deviation of the value x, from the population mean, μ

$(x\text{-}\mu)^2$ is a squared deviation from the mean

$\sum(x\text{-}\mu)^2$ is the sum of the squared deviations

N is the population size

Compare each part of the formula with the steps used to calculate the standard deviation.

» Find the mean of all the data.

» Find the difference between each data point and the mean.

» Square each of the differences.

» Find the mean of the squared differences.

» Take the square root of the mean of the squared differences.

Point out when using the standard deviation application, the calculator or computer software often calculates two slightly different values for the standard deviation, sigma σ or the letter s. σ is the symbol for the population standard deviation and is found using the formula above. The s is the sample standard deviation and is found using the formula:

$$s = \sqrt{\frac{\sum (x-\bar{x})^2}{n-1}}$$

2. When asked what function to use to eliminate the negative signs for the negative deviations, students will often suggest using the absolute value function. A measure of variability that uses the absolute value function is called the mean absolute deviation or MAD. The MAD is interpreted in much the same way as standard deviation.

» Find the mean and the deviations from the mean.

» Take the absolute value of the deviations.

» Find the mean of the absolute deviations.

The formula for MAD is: $\frac{\sum|x-\bar{x}|}{n}$

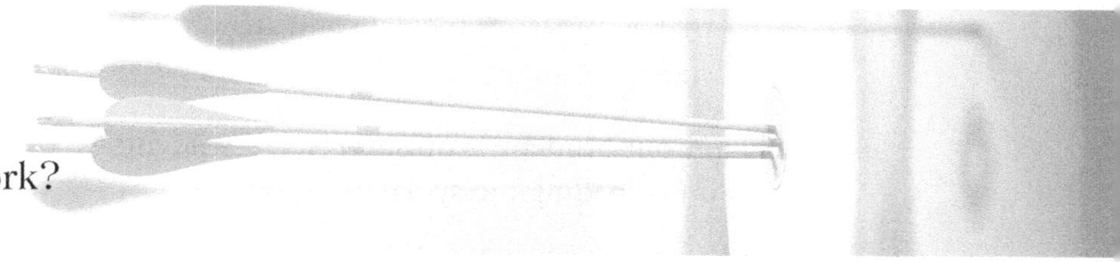

Investigation 4

Do You Have Too Much Homework?

Exploratory Lesson

Overview

This investigation is an open-ended lesson designed to provide an opportunity for students to apply the four components of statistical problem solving. Students use the Random Sampler on the Census at School website and collect data on the number of hours fourth, eighth, and 12th grade students spend per week doing homework. (See Teacher Resource Section at the end of the book for more information about Census at School). After the data are collected, your students will have the opportunity to summarize the data using the techniques studied in investigations 1 to 3. The summaries could include written and oral presentations and/or construction of a poster to display the data and answer the statistical question.

This investigation follows the four components of statistical problem solving put forth in the *Guidelines for Assessment and Instruction in Statistics Education (GAISE) Report*. The four components are formulate a question, design and implement a plan to collect data, analyze the data by measures and graphs, and interpret the results in the context of the original question. This is a GAISE Level B activity.

Instructional Plan

Note: Census at School New Zealand hosts the Random Sampler (*http://new.censusatschool.org.nz/tools/random-sampler*) for international, New Zealand, and cleaned up US data.

The American Statistical Association (ASA) hosts the "messy" US data (*ww2.amstat.org/CensusAtSchool*).

Use the New Zealand web address if you wish to have your students work with data that have been cleaned up (i.e., this database does not have missing data or data that have been entered incorrectly or collected using incorrect units).

Note: If you would like your students to work with "messy" data to provide them the opportunity to investigate "messy" data, then use the Random Sampler on the ASA website. "Messy" data means there are missing data, data collected incorrectly, or data collected using incorrect units.

This lesson is written using the "cleaned up" data from the New Zealand site.

Directions for using messy data on the ASA site can be found on Student Worksheet 4.1.

Brief Overview

» Develop a statistical question pertaining to the number of minutes fourth, eighth, and 12th grade students spend doing homework.

» Use the Census at School Random Sampler to collect data on the number of minutes fourth, eighth, and 12th grade students spent per week doing homework.

» Analyze the data and write a report, create a poster, and/or give an oral report that summarizes the data and answers the statistical question.

Hand out Student Worksheet 4.1 if students are downloading messy data.

Hand out Student Worksheet 4.2 if students are downloading cleaned up data.

Place students into groups of three. Have your students read the following scenario.

Scenario

The National Education Association (NEA) reported that survey data and anecdotal evidence show some students spend many hours nightly doing homework (*www.nea.org/tools/16938.htm*). According to research from the Brookings Institution and Rand Corporation, this homework overload is not the norm. Their researchers analyzed data from a variety of sources and concluded the majority of US

Learning Goal

Summarize numerical data sets by describing the center (mean and/or median), variability (mean absolute deviation and/or standard deviation), and shape (skewed or mound shaped).

Mathematical Practices Through a Statistical Lens

MP1. Make sense of problems and persevere in solving them.

Statistically proficient students understand how to carry out the four steps of the statistical problem-solving process. Students must persevere through the process, adapting and adjusting each component as needed to arrive at a solution that adequately connects the interpretation of results to the statistical question posed.

Materials

Student worksheets are available at *www.statisticsteacher.org/statistics-teacher-publications/focus.*

- Student Worksheet 4.1 Directions for Using ASA Census at School Website (messy data)
- Student Worksheet 4.2 Directions for Using New Zealand Census at School Website (cleaned up data)
- Access to a computer with internet capability and a spreadsheet application such as Excel
- Statistical software or application capable of finding summary statistics and constructing graphs such as dot plots, box plots, and histograms

Estimated Time

One 50-minute class period to collect data and one class period to write and share a report.

Pre-Knowledge

Students should be able to construct a box plot, dot plot, and histogram using technology.

students spend less than an hour a day on homework, regardless of grade level, and this has held true for most of the past 50 years. In the last 20 years, the amount of homework has increased only in the lower grade levels.

Do you spend more time doing homework now than you did when you were in elementary or middle school? This investigation will look at the amount of time fourth, eighth, and 12th grade students spend on homework each week.

Formulate a Statistical Question

Discuss with your students that they will be investigating how many hours per week students in fourth, eighth, and 12th grade spend doing homework each week. Explain that they will take a random sample of students using the Census at School Random Sampler. These data will be used to investigate the statistical question: "How do the number of hours per week fourth graders, eighth graders, and 12th graders spend on homework compare with each other?"

Collect Appropriate Data

Have your students follow the steps outlined on Worksheet 4.1 or 4.2 to demonstrate how to use the Random Sampler on the Census at School site or the New Zealand site.

Note: The population of this study are the students who were involved in responding to this question (How many hours do you spend per week doing homework?) from the United States. The Random Sampler chooses a sample of US students who responded to the question.

Analyze the Data

After your students have taken a random sample from the three grades and used the software to create graphs and summary statistics for the number of hours of homework for each grade level data, ask them to begin their analysis of each grade level's reported number of hours doing homework. Suggest to them that their analysis should include graphs (dot plots, box plots, and/ or histograms) and calculations describing each grade level's distribution (shape, center, and spread). Using the graphs and calculations, students should then compare the hours of homework for each grade level and prepare a report.

Interpret the Results in the Context of the Original Question

Option 1: Write and orally present a report summarizing your results. Your report and presentation should include the following:

» The statistical question that was investigated

» A description of the population sampled

» A summary of the sampling procedure

» Plots of the collected data

» Analysis and descriptions of the data, using calculations and the plots noting any unusual results

» A statement of conclusion about the statistical question

» Recommendations for any follow-up studies or questions that may be investigated

Option 2: Create a poster and orally present the poster summarizing your results.

A data visualization poster is a display containing two or more related graphics that summarize a set of data, look at the data from different points of view, and answer specific statistical questions about the data.

The poster and presentation should include the following:

» The statistical question that was investigated as the title of the poster

» A description of the population sampled (in the oral report)

» A summary of the sampling procedure (in the oral report)

» The organized collected data—tables and plots (at least two graphs)

» Analysis and descriptions of the data, using calculations and the plots noting any unusual results (in the oral report)

» A statement of conclusion about the statistical question

» Recommendations for any follow-up studies or questions that may be investigated (in the oral report)

Note: A rubric for evaluating the posters can be found at *www.amstat.org*. The rubric is found under the Education tab and then K–12 Educators, Student Competitions. You can also download it from *www.amstat.org/asa/files/pdfs/EDU-PosterJudgingRubric.pdf*.

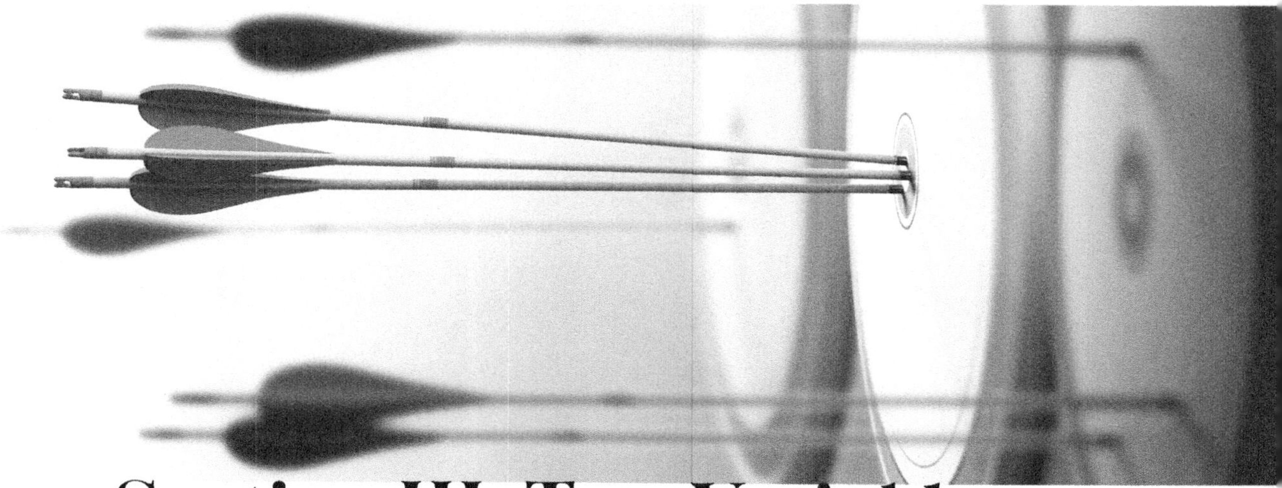

Section III: Two-Variable Data Analysis

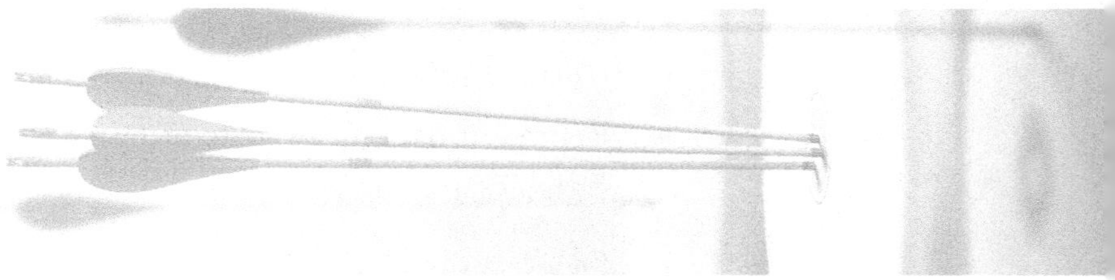

Investigation 5

How Many Calories?

Scatterplots

Overview

This investigation begins the exploration of relationships within bivariate data by investigating errors made by students between guesses and actual values, setting the stage for the concept of a residual.

The concept of a residual and a residual plot will be used in Investigation 7 as a tool for exploring variability about the least squares regression line.

This investigation follows the four components of statistical problem solving put forth in the *Guidelines for Assessment and Instruction in Statistics Education (GAISE) Pre-K-12 Report.* The four components are formulate a statistical question, design and implement a plan to collect data, analyze the data by measures and graphs, and interpret the results in the context of the original question. This is a GAISE Level B activity.

This activity is based on lessons from *Exploring Linear Relations* (Lesson 7), published by the American Statistical Association (original copyright by Dale Seymour Publications 1999). *Exploring Linear Relations* is a module in the ASA Data-Driven Mathematics Project. It is available as a free download at *www.amstat.org/asa/files/pdfs/ddmseries/Exploring-LinearRelations--TeachersEdition.pdf.*

Instructional Plan

Brief Overview

» Read and discuss the scenario about obesity and caloric information.

» Formulate the statistical question: What is the typical error made by students in estimating the number of calories in bite-sized candies?

» Have students estimate the number of calories in each item and find their errors.

» Use the squares of the errors to determine who is the "best" guesser.

Hand out Student Worksheet 5.1 Guess the Calories. Have your students read the Scenario.

Scenario

The following excerpt comes from *Attacking the Obesity Epidemic: The Potential Health Benefits of Providing Nutrition Information in Restaurants* by Scot Burton, Elizabeth H. Creyer, Jeremy Kees, and Kyle Huggins. The entire article can be found at *www.ncbi.nlm.nih.gov/pmc/articles/PMC1551968.*

> Sixty-four percent of American adults are either overweight or obese, and the obesity epidemic shows few signs of weakening. Although the precise number of deaths attributable to obesity is difficult to estimate, obesity is clearly a major cause of preventable death. Not surprisingly, improving the healthfulness of the American diet has become a national health priority. The increasing prevalence of obesity-related diseases has been blamed, in part, on the increased consumption of foods prepared outside the home. Restaurant expenditures have increased consistently in

Learning Goals

» Represent data on two quantitative variables on a scatterplot and describe how the variables are related

» Develop understanding of an error

Mathematical Practices Through a Statistical Lens

MP3. Construct viable arguments and critique the reasoning of others.

Statistically proficient students use appropriate data and statistical methods to draw conclusions about a statistical question. They follow the logical progression of the statistical problem-solving process to investigate answers to a statistical question and provide insights into the research topic. They reason inductively about data, making inferences that consider the context from which the data arose. They justify their conclusions, communicate them to others (orally and in writing), and critique the conclusions of others.

Materials

Student worksheets are available at *www.statisticsteacher.org/statistics-teacher-publications/focus.*

» Student Worksheet 5.1 Guess the Calories

» Graph paper (.5 cm or .25 inch)

» "Fun" size Milky Way candy bar

Estimated Time

One 50-minute class period

Pre-Knowledge

Students should already be able to:

» Construct scatterplots

» Draw the y=x line on a scatterplot

recent decades; consumers now spend more than $400 billion annually.

Results: Survey results showed that levels of calories, fat, and saturated fat in less-healthful restaurant items were significantly underestimated by consumers. Actual fat and saturated fat levels were twice consumers' estimates and calories approached two times more than what consumers expected. In the subsequent experiment, for items for which levels of calories, fat, and saturated fat substantially exceeded consumers' expectations, the provision of nutrition information had a significant influence on product attitude, purchase intention, and choice.

Conclusions: Most consumers are unaware of the high levels of calories, fat, saturated fat, and sodium found in many menu items. Provision of nutrition information on restaurant menus could potentially have a positive impact on public health by reducing the consumption of less-healthful foods.

Formulate a Statistical Question

After students have read the scenario ask:

How well do you think you might estimate the calories in restaurant items? Have you noticed restaurants providing this information more readily? For example, some restaurants list this on their menus, such as Panera and Noodles and Co.

Explain that they will be asked to estimate the number of calories in "fun" size candy bars. Hold up a fun size Milky Way candy bar and ask students to guess the number of calories. Have them write down their guesses, and then ask a few students to share. Then share that the fun size Milky Way bar contains 80 calories.

Ask your students how close their estimates were for the number of calories.

Ask students to create a possible statistical question for this activity.

Example: *What is the typical error made by students in estimating the number of calories in bite-sized candies?*

Note: If it appears students may interpret this investigation as the worst guessers will become obese, it may be worth making the point that the study was looking for a positive relationship between knowing the calories of a food prior to consumption and making healthier choices. This does not indicate not knowing the calories in food means making unhealthy choices.

Table 5.1 Candy

Candy Item – Fun Size	Actual
Snickers	80
Skittles	80
Butterfinger	100
Kit Kat	70
M&M's Plain	73
M&M's Peanut	90
Reese's Peanut Butter Cup	110
Starburst	40
Whoppers	100
Twizzlers	50
Jolly Ranchers (3 Pieces)	70

Collect Appropriate Data

Ask students to complete problem 1 on Student Worksheet 5.1 Guess the Calories.

1. Fill in the "Guess?" column with your guesses for the number of calories in each fun size candy item.

When students have completed their Guess? column, reveal the actual number of calories (Table 5.1). Have the students complete Problem 2.

2. Fill in the "Actual" column with the actual number of calories in each fun size candy item.

Analyze the Data

Ask your students to complete Question 3.

3. How might you decide who is the best guesser in the class? Justify your answer.

After giving your students time to individually write about their methods of determining the best guesser, have them share with a partner or in small groups. Then, as a whole group, ask students to share their ideas about how they might determine who is the best guesser. Does a person need to be exactly correct? Or just close? Is underestimating different from overestimating?

Possible answers: *Students might suggest a variety of ways to calculate who is the best guesser. Encourage discussion and collect several ideas. For example, the greatest number of guesses that match actual calories or the greatest number of guesses that were within 10 calories of the actual calories. If it doesn't come up, guide discussion around whether it matters how far off someone was.*

Explain that we are going to create a visual display of their data to help determine who is the best guesser. Distribute graph paper to each student. Give students time to answer questions 4 and 5.

4. Create a scatterplot of your data on graph paper, plotting the actual calories on the x-axis and your guess on the y-axis.

Sample answer: *Figure 5.1*

5. Describe the relationship between your guesses and the actual calories in each candy item.

Possible answer: *As the actual calories increase, my guessed calories increased. I tended to underestimate the actual calories. For the Butterfinger,*

Table 5.2 Sample Data

Candy Item – Fun Size	Actual	Guess?
Snickers	80	90
Skittles	80	75
Butterfinger	100	150
Kit Kat	70	50
M&M's Plain	73	50
M&M's Peanut	90	85
Reese's Peanut Butter Cup	110	90
Starburst	40	25
Whoppers	100	50
Twizzlers	50	30
Jolly Ranchers (3 Pieces)	70	50

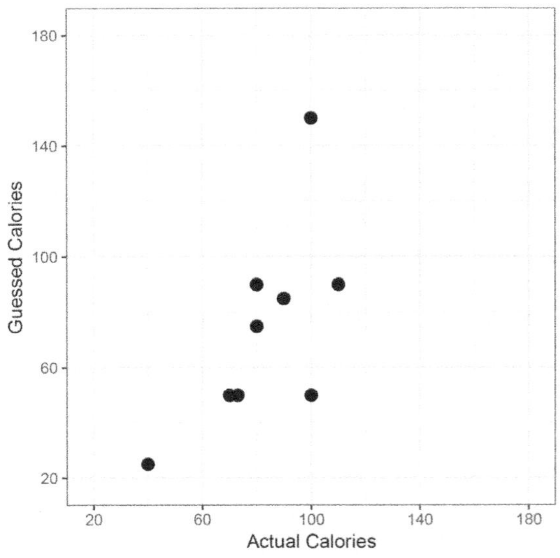

Figure 5.1: Scatterplot of guessed and actual calories

my guess for number of calories was much higher than the actual number of calories.

Use one of the student's scatterplots as an example or show students a scatterplot with the x-axis labeled Actual Calories and the y-axis labeled Guessed Calories. Ask what a point on this graph represents. For example, the point (100, 75) represents a candy item that has 100 calories and was guessed to have 75 calories. Refer to the study from the beginning of the lesson—if someone often underestimated the number of calories in the candy items, how would that appear in a scatterplot of the data?

The points would be closer to the x-axis since the guessed calories would be lower than the actual calories.

Ask your students to answer Question 6,

6. What would the scatterplot look like if someone had guessed the correct actual calories in each candy item?

Possible answer: *The points would make a straight line. If it doesn't come up, ask for the equation of the line, which is $y = x$, the line* *where all the y-values (guessed calories) are the same as the x-values (actual calories).*

Have students draw the y = x line on their scatterplots. Ask what it means for a point to be on the line, below the line, and above the line.

Possible answers: *If a point is on the line, the guess matches the actual calories. If a point is below the line, the guess is lower than the actual calories. If a point is above the line, the guess is higher than the actual calories.*

Ask students to answer Question 7, and then have some students share their scatterplots and explain the type of guesser their graph shows.

7. Describe the type of guesser your scatterplot shows. Explain.

Possible answer: *Based on my scatterplot (Figure 5.2), I tended to underestimate the calories in the fun size items as shown by most of my dots being below the line y = x.*

Ask students how this line might help us determine who is a better guesser.

Possible answer: *Students might suggest a person whose points are closer to the line is a better guesser.*

How might we measure the distance from the line?

Possible answer: *Students might suggest measuring the perpendicular distances, horizontal distances, vertical distances, or other methods.*

Explain that we are going to use the vertical distances, as this is a convention used in statistics. What do the vertical distances represent?

Possible answer: *The difference between the guessed calories (y-value of the point) and the actual calories (y-value on the line).*

What could we call these differences?

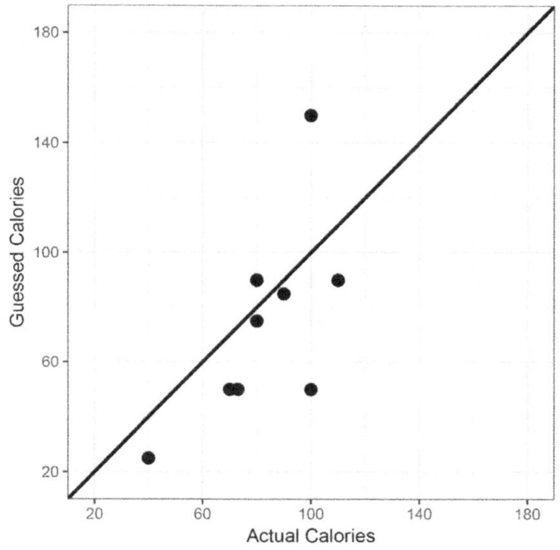

Figure 5.2: Scatterplot of someone who underestimated the calories in the candy.

Possible answer: *These differences between the guessed calories and actual calories for each candy item are the amounts of error for each candy item.*

Note: You should not use the term "residual." This activity is setting the stage for understanding the concept of the residual in Investigation 7.

The values for the vertical distances represent the error for each guess. Using a student's scatterplot, choose a point above the line and draw the vertical distance between the point and the line y = x. The example here shows the point (100, 150), which means the difference between the guessed calories (150) minus the actual calories (100) is 50, thus the error is 50 calories.

If an error is 10, how could we determine if the person overestimated or underestimated? If no students make a suggestion, explain that positives and negatives are used to determine this. The error is positive if the point is above the line, and the error is negative if the point is below the line. In this situation, points above

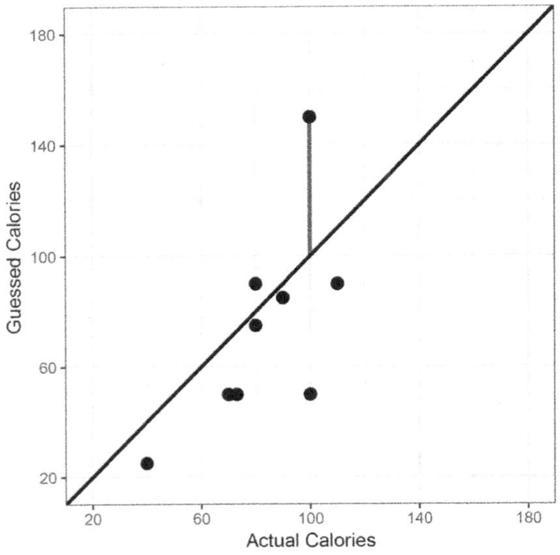

Figure 5.3: Scatterplot showing the difference between the guessed calories and actual calories for each candy

the line are overestimates and positive errors; points below the line are underestimates and negative errors. It might be helpful to show another example on a student's graph.

Ask what an error of 0 represents.

Answer: *The guess and the actual value were the same.*

Have your students answer questions 8 and 9.

8. On your scatterplot, draw the y=x line. Then draw the vertical distances representing the "errors" on your scatterplot.
9. On the table of guesses and actual number of calories, add a third column labeled "Errors" and calculate the errors (guess minus actual) for each candy item. Find the sum of the errors.

Answer based on the given example (Table 5.3).

Ask students how we might use the errors to help us determine who is the best guesser. Encourage discussion. Students might suggest how many are close to 0 or +/- 10 calories or adding up the errors. Propose that the best

Table 5.3 Errors Based on Sample Data

Candy Item – Fun Size	Actual	Guess?	Errors
Snickers	80	90	10
Skittles	80	75	-5
Butterfinger	100	150	50
Kit Kat	70	50	-20
M&M's Plain	73	50	-23
M&M's Peanut	90	85	-5
Reese's Peanut Butter Cup	110	90	-20
Starburst	40	25	-15
Whoppers	100	50	-50
Twizzlers	50	30	-20
Jolly Ranchers (3 Pieces)	70	50	-20

guesser is one who has a sum of errors close to zero. After students have found their sum, ask who the best guesser was using this method. Ask students if they have any concerns about using the sum of the errors.

Possible answer: *This would not be a good method since someone could have a very high error (extreme overestimate) balanced by a very low error (extreme underestimate).*

Ask students how we might handle values that are negative when we might like them to be positive.

Possible answer: *Students will most likely say take the absolute value.*

Have your students answer Question 10.

10. On the table of guesses and actual number of calories, add a fourth column labeled "Absolute Value" and complete the column. Find the sum of the absolute values.

Note: Finding the absolute values and the sum of the absolute values could be connected to earlier learning of the Mean Absolute Deviation (MAD) from middle school.

Now ask again who is the best guesser. The best guesser would have the lowest sum of the absolute value of errors.

Ask students for another way we might handle values that are negative when we might like them to be positive.

Possible answer: *We could square the values.*

Ask students where else squaring has been used in statistics to "handle" negatives.

Answer: *Squaring was used when calculating deviations from the mean for standard deviation.*

Ask how this would help determine who the best guesser is.

Possible answer: *The best guesser would have the lowest sum of the squares of the errors, though this will usually be much higher than the sum of the absolute value of the errors.*

Note: If all the errors are less than 1, then the sum of squares will be less.

Have your students answer Question 11.

11. On the table of guesses and actual number of calories, add a fifth column labeled "Squares" and complete the column. Find the sum of the squares.

Ask again who the best guesser is. Did this answer change from the best guesser based on absolute values of the errors?

Interpret the Results in the Context of the Original Question

In groups of four, ask your students to complete problems 12 to 15. Then discuss.

12. Compare your results from Question 11 with the other students in your group. Who in your group was the best guesser of calories? Justify your answer.

Possible answer: *I know ____ is the best guesser because his/her sum of the squared errors is the lowest. This means his/her guesses were closest to the line y=x, which would be the line created if all the guesses were correct.*

13. Using the scatterplots and analysis of the errors, answer the statistical question: What is the typical error made by students in estimating the number of calories in bite-sized candies?

Answers will vary depending on the class results. Reference whether the students tended to overestimate or underestimate. Also refer to the usual number of calories their estimate was off.

14. How do these results relate to the study results?

Possible answer: *Based on the results from the class, either support or do not support the claim in the study that people tend to underestimate the number of calories in restaurant items.*

15. Why might finding errors be important when looking at data?

Possible answer: *Errors can help determine how close guesses are to the actual data values and who is the best guesser from many guessers.*

Additional Ideas

Do the same investigation steps, but instead of using the candy items, guess the calories (or fat grams) of fast food items from a particular restaurant or several restaurants. Students can find this information online.

Do the same investigation steps, but instead of using the candy items, guess the ages of various celebrities such as actors, sports figures, or politicians (international, national, and/or local) students would know. It can be helpful to have recent pictures of the celebrities to show. Make sure to have variety in ages.

Exit Ticket

1. The Price Is Right - Cliffhanger Game: "The contestant bids on three small prizes. For every dollar the contestant is away from the actual prices, a mountain climber takes one step up a mountain. If the mountain climber does not exceed 25 steps after the contestant has bid on all three prizes, then the contestant wins a bonus prize."
   Source: www.priceisright.com/games

Here are some items and the prices the contestants bid. Who won the bonus prize? Who was the best price guesser? Who was the worst guesser? Justify your answer.

Jenna		
Item	**Bid**	**Actual**
Passport Holder	12	16
Toaster	35	22
Coffee Pot	35	40
Lindsey		
Item	**Bid**	**Actual**
Inflatable Pool Lounge Chair	15	20
Electronic Piggy Bank	37	30
Pet Brush and Accessories	27	40
Debra		
Item	**Bid**	**Actual**
Liquid Measuring Cup	5	15
Electric Egg Cooker	7	22
Whipped Cream Dispenser	6	26
Kristen		
Item	**Bid**	**Actual**
Steam Iron	25	22
Electric Heater Fan	35	35
Milkshake Drink Mixer	50	49

Exit Ticket Cont.

Exit Ticket Answer:

Jenna					
Item	**Bid**	**Actual**	**Error**	**Absolute Value of Error**	**Squared Error**
Passport Holder	12	16	-4	4	16
Toaster	35	22	13	13	169
Coffee Pot	35	40	-5	5	25
Sum	22	210			
Lindsey					
Item	**Bid**	**Actual**	**Error**	**Absolute Value of Error**	**Squared Error**
Inflatable Pool Lounge Chair	15	20	-5	5	25
Electronic Piggy Bank	37	30	7	7	49
Pet Brush and Accessories	27	40	-13	13	169
Sum	25	243			
Debra					
Item	**Bid**	**Actual**	**Error**	**Absolute Value of Error**	**Squared Error**
Liquid Measuring Cup	5	15	-10	10	100
Electric Egg Cooker	7	22	-15	15	225
Whipped Cream Dispenser	6	26	-20	20	400
Sum	45	725			
Kristen					
Item	**Bid**	**Actual**	**Error**	**Absolute Value of Error**	**Squared Error**
Steam Iron	25	22	3	3	9
Electric Heater Fan	35	35	0	0	0
Milkshake Drink Mixer	50	49	1	1	1
Sum	4	10			

Jenna, Lindsey, and Kristen all won the bonus prize because their sum of the absolute values of the errors did not exceed 25. The best guesser is Kristen, since her sum of squared errors is only 10. The worst guesser is Debra, as her sum of squared errors is 725. Debra underestimated all the prices by quite a bit.

Further Explorations and Extensions

1. The Price Is Right - Bargain Game: "Two prizes are shown to the contestant. Each prize displays a bargain price that is below its actual retail price. If the contestant selects the bigger bargain (the prize with the bargain price that is farther from the actual retail price), then the contestant wins both prizes."
   Source: www.priceisright.com/games

Sketch a scatterplot that would depict what several prizes and their bargain prices might look like, as well as the line y = x.

Example video: *www.youtube.com/watch?v=crzYmKNvISg*

Example from video: Which price is the bigger bargain? Billiards table at $1600 or 55" HDTV at $2649?

Billiards table: bargain price: $1600; actual price: $2600

55" HDTV: bargain price: $2649; actual price: $3149

***Answers:** Figure 5.4 and Figure 5.5*

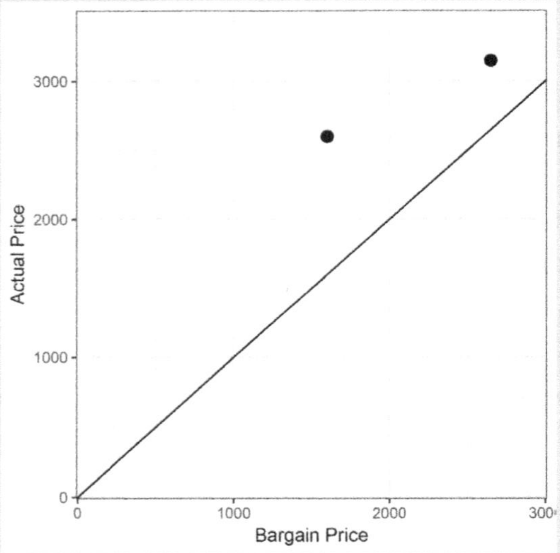

Figure 5.4

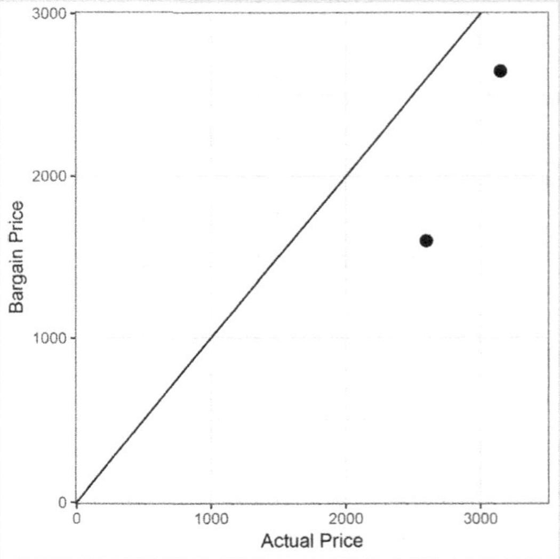

Figure 5.5

Investigation 6

Are Gender and Pay Related?
Correlation

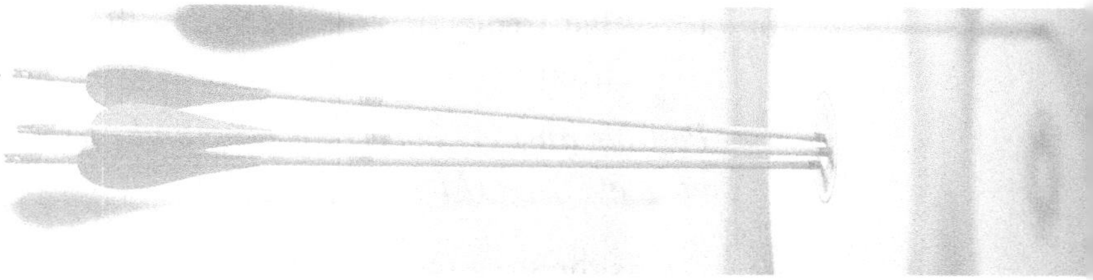

Overview

This investigation continues to explore relationships between two quantitative variables. It focuses on determining whether there is an association between two quantitative variables by looking for patterns in scatterplots and describing the relationship between two quantitative variables by finding and interpreting the correlation coefficient.

The next two investigations focus on these relationships by exploring the variability about the least squares regression line using two tools: correlation coefficient and residual plots.

This investigation follows the four components of statistical problem solving put forth in the *Guidelines for Assessment and Instruction in Statistics Education (GAISE) Report*. The four components are formulate a question, design and implement a plan to collect data, analyze the data by measures and graphs, and interpret the results in the context of the original question. This is a GAISE Level B activity.

Instructional Plan

Brief Overview

Part I: Introduce the three types of association.

Part II: Develop the concept of the correlation coefficient.

Part III: Interpret the correlation coefficient in context.

Part I: Association

Hand out Student Worksheet 6.1 Patterns in Scatterplots. Ask your students to answer questions 1 and 2.

Do you own an MP3 player? Or do you have a lot of songs stored on your cell phone? How much memory are the songs using on the device?

Do you think there is a relationship between the length of a song in minutes and the size of the file on an MP3 player?

The scatterplot in Figure 6.1 shows the file sizes (in MB) for songs of different lengths (in seconds).

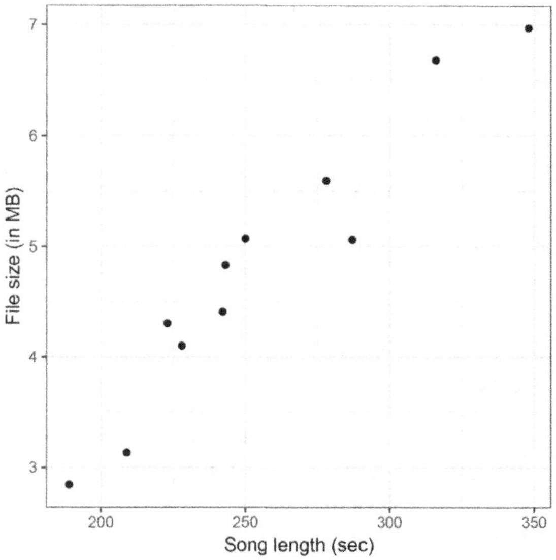

Figure 6.1: File sizes for songs of different lengths

1. What trends do you observe in the scatterplot?

Possible answer: *Longer songs (seconds) tend to use more space (MB).*

2. As the song increases in length, what is happening to the size of the file?

Possible answer: *The file size is increasing.*

Explain that, on this scatterplot, the length of the song is called the *independent variable* or *explanatory variable* and the size of the file is called the *dependent variable* or *response variable.*

Independent Variable or Explanatory Variable
The explanatory variables are those variables that influence changes in the response variable.

Learning Goal

Represent data of two quantitative variables on a scatterplot and discuss whether the two variables are related. Compute (using technology) and interpret the correlation coefficient of a linear fit.

Mathematical Practices Through a Statistical Lens

MP7. Look for and make use of structure.

Students use structure to separate the 'signal' from the 'noise' in a set of data—the 'signal' being the structure, the 'noise' being the variability. They look for patterns in the variability around the structure and recognize that these patterns can often be quantified.

Materials

Student worksheets are available at *www.statisticsteacher.org/statistics-teacher-publications/focus.*

» Statistical technology that can do the following:

- Create a scatterplot
- Calculate Pearson's correlation coefficient

» Student Worksheet 6.1 Patterns in Scatterplots

» Student Worksheet 6.2 Gender and Pay

» Student Worksheet 6.3 Graphs to Illustrate Association

» Exit Ticket

Estimated Time

Two 50-minute class periods: One period for introducing correlation (parts I and II) and another period to apply the concept of correlation (Part III)

Pre-Knowledge

Students should be able to construct scatterplots using technology.

Dependent Variable or Response Variable
The response variable is the observed result of the explanatory variable being manipulated.

When describing the scatterplot, statisticians generally list the dependent variable (located on the y-axis) vs. the independent variable (located on the x-axis). For example, statisticians would describe the previous scatterplot as file size vs. length of song.

There are three types of association when describing a scatterplot: positive association, negative association, and no association.

Note: Figure 6.2 (Flight time vs. Distance), Figure 6.3 (Olympic 100 m freestyle time vs. Years since 1900), and Figure 6.4 (Gross earnings vs. movie running time) can be found on Student Worksheet 6.3 Graphs to Illustrate Association.

Display Figure 6.2, an example of a scatterplot showing a **positive association.**

The data show the distance in miles and time in minutes for a sample of United Airline nonstop flights from Chicago to various cities west of Chicago.

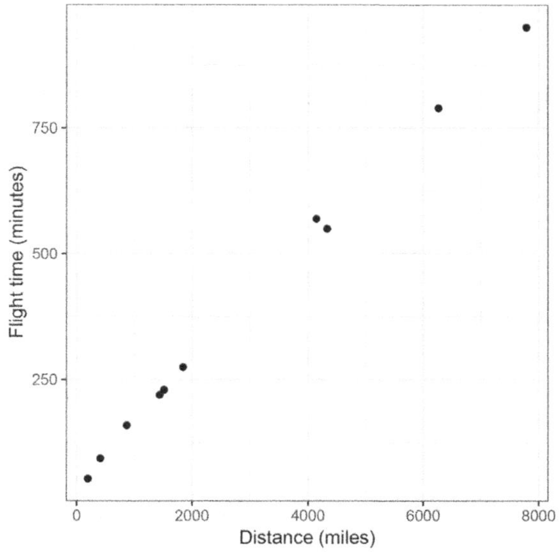

Figure 6.2: A scatterplot showing a positive association

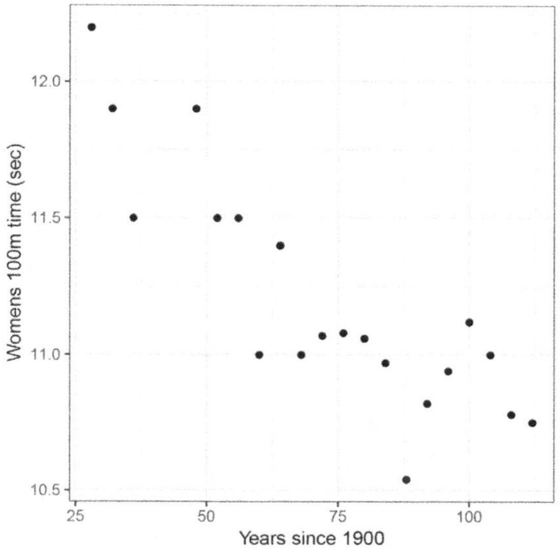

Figure 6.3: A scatterplot showing a negative association

A positive association occurs when large values of one variable are associated with large values of the other and small with small.

Or

As the distance from Chicago increases, the flight time increases.

Display Figure 6.3, an example of a scatterplot showing a **negative association.**

The data show Olympic Women's 100 m freestyle times since 1912.

A negative association occurs when the values of one variable tend to decrease as the values of the other variable increase.

Or

As the years have increased from 1912, the Olympic times in the women's 100 m freestyle have decreased.

Display Figure 6.4, an example of a scatterplot showing **no association.**

Data are the running times (minutes) and gross receipts for a selected group of movies.

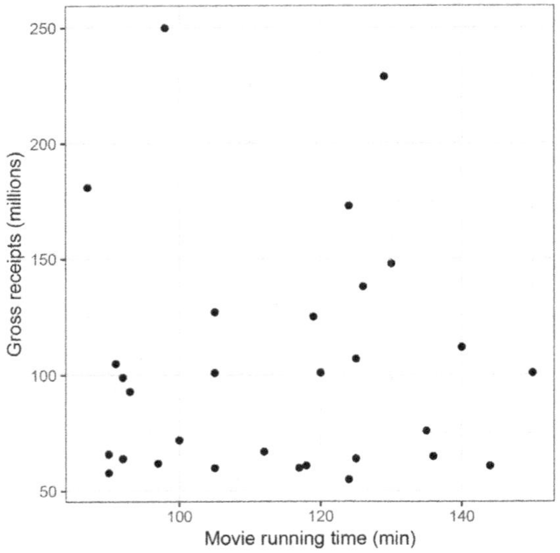

Figure 6.4: A scatterplot showing no association

No association occurs when one variable increases and there is no pattern for how the other variable reacts.

Or

As the running time of a movie increased, the amount of money the movie grossed was hard to predict—sometimes it was large, sometimes it remained fairly constant.

Ask your students to answer Question 3.

3. How would you describe the relationship between the length of a song and the size of the file?

Answer: *There is a positive association. As the song length increases, the file size increases.*

This is a short activity to help solidify student understanding of association. Ask students to stand up. Explain that sets of variables will be shared and students should raise both arms if they think there is a positive correlation between the variables, put their arms straight out if they think there is no correlation, and put their arms down at their sides if they think there is a negative correlation. (These variables could be altered to reflect the interests of your students.)

» Reading achievement vs. IQ
Positive Association

» Test scores vs. level of anxiety
Negative Association

» Weight gain vs. amount of exercise
Negative Association

» Math achievement vs. reading achievement
Positive Association

» Math achievement vs. athletic achievement
No Association

» Lifetime earnings vs. years of schooling
Positive Association

» Number of texts sent/received in a day vs. number of Facebook friends
Could be arguments for all three answers—most likely No Association or Positive Association

Have a brief discussion about the amount of time it took students to position their arms. Some of the sets of variables may have been easier than others for students to make decisions. The amount of time to decide about correlation is related to the strength of the correlation.

Note: It is important to point out to your students that association does not mean causation. A positive association between reading achievement and math achievement does not mean high reading achievement causes high math achievement.

Part II: Correlation Coefficient

Explain that our goal is to find a mathematical model to describe the relationship between two quantitative variables. For example, what mathematical model could be used

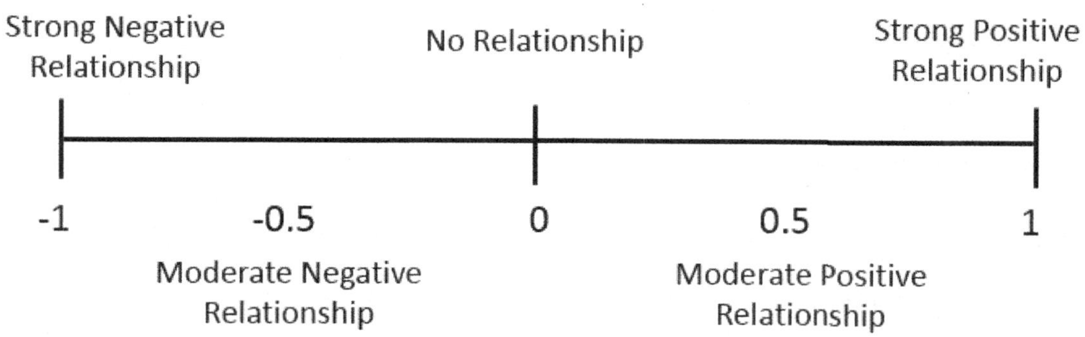

to describe the association between the length of a song and its file size?

Answer: *A linear model makes sense because the data appear to fit close to a line.*

If the association appears linear, then the "tightness" of the data points to the line fitted to the data can be measured. This value is called the correlation coefficient.

A correlation coefficient is a number that measures the direction and strength of a linear relationship between two quantitative variables. One such measure is called Pearson's correlation coefficient, represented by the letter *r*.

Note: Explain that the convention is to call this value "r." The letter r was used because Sir Francis Galton, who developed the concept of regression, originally used r to describe the slope of the regression line. While Pearson developed the formula we use today, the use of r stuck with correlation coefficient. *Source: www.buttelake.com/corr.htm*

Explain that the closer Pearson's correlation coefficient is to +1, the stronger the positive linear relationship. The closer Pearson's correlation coefficient is to -1, the stronger the negative linear relationship.

Display the diagram at the top of the page.

Note: For more information about the development of the formula for Pearson's correlation coefficient, see the Further Explorations and Extensions section at the end of this investigation.

The next step is to have students estimate the correlation coefficient for the following four scatterplots. The plots are the same graphs used to introduce association. They are found on Worksheet 6.3 Graphs to Illustrate Association.

Note: Students could be asked to place themselves along an imaginary number line in the room to estimate the correlation coefficient.

Display the scatterplot of flight time versus distance from Chicago again and ask students to estimate what the correlation coefficient might be.

Answer: *The correlation coefficient is 0.999.*

Display the scatterplot of the Olympic times versus years since 1900 again and ask students to estimate what the correlation coefficient might be.

Answer: *The correlation coefficient is -0.95.*

Display the scatterplot of the gross receipts versus movie run times again and ask students to estimate what the correlation coefficient might be.

Answer: *The correlation coefficient is 0.01.*

Display the scatterplot of the length of song and file size versus length of song again and ask students to estimate what the correlation coefficient might be.

Answer: *The correlation coefficient is 0.96.*

Optional: The websites shown below can provide extra practice for students to get a good feel for what different values of r look like.

www.istics.net/Correlations: Students are presented four scatterplots and asked to match each correlation coefficient to the correct graph.

http://guessthecorrelation.com: A game in which students guess the correlation and compete for high scores. Students can also play against each other by entering the competitor's user name.

Part III: Investigating a Scenario

Note: This part of the lesson is based on data from *www.census.gov*, obtained through Core Math Tools at *www.nctm.org/Classroom-Resources/Core-Math-Tools/Core-Math-Tools.* Another website that provides quite a bit of information, including a breakdown by state, is the American Association of University Women at *www.aauw.org/research/the-simple-truth-about-the-gender-pay-gap*. Students might be interested in doing more research on this topic.

Hand out Student Worksheet 6.2 Gender and Pay. Have students read the scenario.

Scenario

> Did you know that in 2016, women working full time in the United States typically were paid just 80 percent of what men were paid, a gap of 20 percent? The gap has narrowed since the 1970s, due largely to women's progress in education and workforce participation and to men's wages rising at a slower rate. Still, the pay gap does not appear likely to go away on its own. At the rate of change between 1960 and 2016, women are expected to reach pay equity with men in 2059. But even that slow progress has stalled in recent years. If change continues at the slower rate seen since 2001, women will not reach pay equity with men until 2119.
> *Source: www.aauw.org/research/the-simple-truth-about-the-gender-pay-gap*

Note: Another interesting article for students to explore the gender pay gap is by Amanda Golbeck and titled "How One Woman Used Regression to Influence the Salaries of Many." It tells the story of Elizabeth Scott, the Berkeley statistics professor who spent two decades analyzing inequities in academic salaries and advocating for change. Find it at *http://onlinelibrary.wiley.com/doi/10.1111/j.1740-9713.2017.01092.x/epdf.*

After students have read the scenario, ask them to brainstorm about why this gap might exist. Ideas that might surface include type of job, education level, experience, age, and race. Some students might debate whether the pay gap between gender exists, as some claim the 20% is inflated. Some sites to explore are the following:

» *www.huffingtonpost.com/christina-hoff-sommers/wage-gap_b_2073804.html*

» *www.bls.gov/opub/reports/womens-earnings/archive/womensearnings_2009.pdf*

Formulate a Statistical Question

Point out to students that, in the scenario, the gap was measured as a percent, calculating the percent women were paid compared to men. This is often referred to as women's to men's earnings ratio. For this investigation, this value will simply be referred to as the "earnings ratio."

After the discussion concerning the pay gap, discuss possible statistical questions students could investigate. To help the discussion, remind them they need to identify the populations of interest and how these populations could be compared.

Table 6.1 Answers to Question 1

Years since 1970	Median Income Men ($)	Median Income Women ($)	Earnings Ratio
0	9,184	5,440	0.592
5	12,934	7,719	0.597
10	19,173	11,591	0.605
15	24,999	16,252	0.650
20	28,979	20,591	0.711
25	32,199	23,777	0.738
30	38,891	29,123	0.749
35	42,188	33,256	0.788
39	49,164	37,234	0.757
45	50,119	40,022	0.799

On Student Worksheet 6.2 Gender and Pay, ask your students to write a possible statistical question. Have them compare with others in their group.

In this investigation, the statistical question we will be investigating is: "To what extent are the median income of US males and the median income of US females related?"

Collection of Data

Explain how the earnings ratio is calculated for 1970 (0 years since 1970).

Ask your students to interpret the earnings ratio in 1970.

Possible answer: *The earnings ratio is 5440/9184, also written as approximately 0.592, which means women earned about $0.59 for every $1 men earned in 1970.*

Analyze the Data

Ask your students to complete questions 1 to 4 on the student worksheet.

1. Find the remaining earnings ratios.

See answers in Table 6.1.

2. Interpret the earnings ratio for 45 years since 1970.

Possible answer: *Women earned about $0.80 for every $1 men earned in 2015.*

3. What trends do you observe in the earnings ratio over time?

Possible answer: *The earnings ratio has increased since 1970.*

4. Could the earnings ratio exceed 1? What would that mean?

Possible answer: *The earnings ratio would exceed 1 when the median income for women exceeds the median income for men.*

Explain that the response variable will be the earnings ratio, as this amount is changing over time. Time (years since 1970) is the explanatory variable, as it is possibly influencing the change in the earnings ratio.

Ask your students to complete Question 5 on the student worksheet.

5. Use technology to create a scatterplot of time and earnings ratio. Sketch a copy of the scatterplot.

Answer: Figure 6.5.

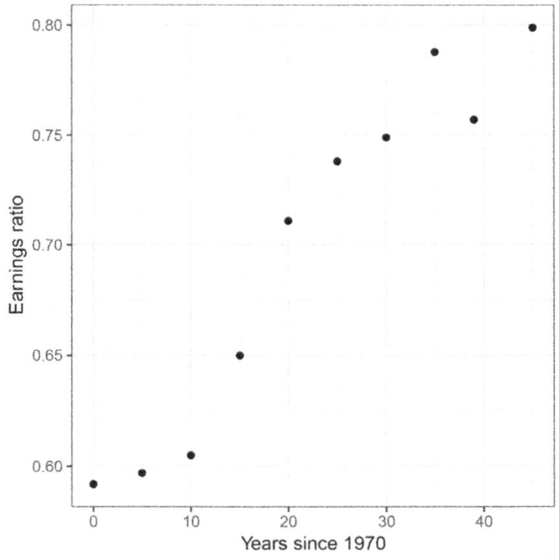

Figure 6.5: A scatterplot of time and earnings ratio

Ask students to complete questions 6 to 8 on the student worksheet.

6. Describe the relationship between the earnings ratio and time since 1970.

Possible answer: *There is a positive association. As the years since 1970 increases, the earnings ratio increases. In other words, the ratio of the median earnings for women to the median earnings for men is getting larger, thus the gap between pay based on gender is diminishing over time.*

7. Estimate r, the correlation coefficient.

Possible answer: *Approximately 0.95.*

8. Do you think it would be appropriate to draw a line through the data?

Possible answer: *Although there is some curvature, it appears a linear function could be used to model the relationship between earnings ratio and time since 1970. The fit should be relatively strong.*

Explain that the correlation coefficient can easily be found using technology. Graphing calculators, spreadsheets, and statistical software/applications all have the capability of finding the value of r. Demonstrate how to find the value of r. Ask students to complete Question 9.

9. Use technology to find the value of r and interpret this value in terms of the data.

Possible answer: *r is approximately 0.96. The data fit very tightly to a line. The earnings ratio and time have a strong positive correlation.*

Interpret the Results in the Context of the Original Question

Ask students to revisit or restate the statistical question: "To what extent are the median income of males and the median income of females related?"

Ask students to discuss with a partner or small group how they would answer the question based on the scatterplot and correlation coefficient, and then write a few sentences to answer the statistical question based on the analysis.

Optional: Ask students to comment on possible social implications of the results or other data they might be interested in collecting that might influence the answer to the statistical question.

Example: A scatterplot of the data shows a positive linear pattern. The correlation coefficient is approximately 0.963, indicating a strong positive correlation between the years since 1970 and the earnings ratio of median incomes of women to men. If this pattern continues, the earnings ratio will eventually be 1, indicating no gender gap in income.

Additional Ideas

This lesson could focus on a different source of data, such as the cost of a 30-second Super Bowl ad and the winning team player's share (data set available through Core Math Tools).

Exit Ticket

For the following four scatterplots:

» Summarize the relationship between the variables with a sentence.

» Determine if a linear approximation is appropriate. If so, estimate the correlation coefficient.

All data are sourced from Core Math Tools, available from *www.nctm.org/Classroom-Resources/Core-Math-Tools/Core-Math-Tools.*

1. World Series Average vs. Regular Season Average

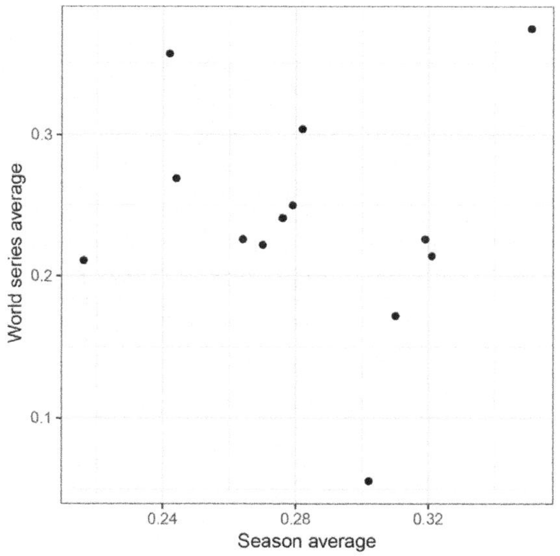

Possible answer: *If a linear relationship exists, it would appear to be weak and maybe negative, indicating there is little to no linear relationship between a player's World Series batting average and regular season batting average. The actual value for r = -0.03.*

2. Highway mpg vs. Curb Weight for Selected Cars

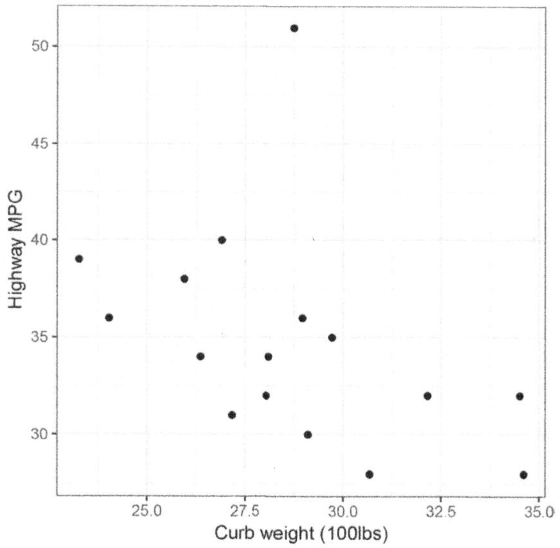

Possible answer: *It appears there might be a moderate to weak negative correlation between highway mpg and curb weight in pounds for selected compact cars. The actual value for r = -0.43.*

Exit Ticket Cont.

3. Age (Months) Baby Began to Crawl vs. Average Outside Temperature

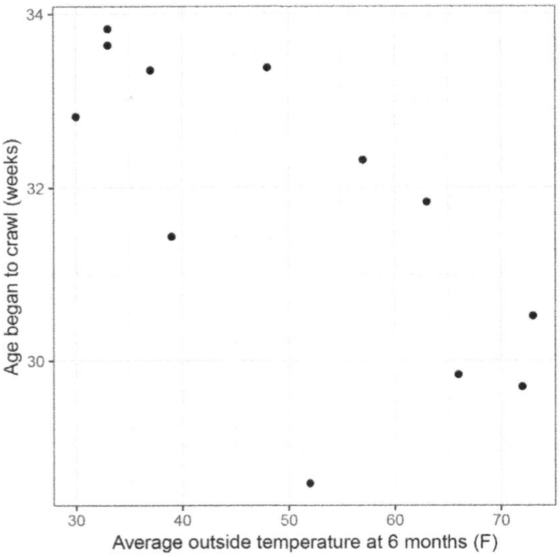

Possible answer: *It appears there is a moderate negative linear relationship between the age babies began to crawl and the average outside temperature. The actual value for r = -0.7.*

4. Distance (feet) Until Car Stopped vs. Speed (mph)

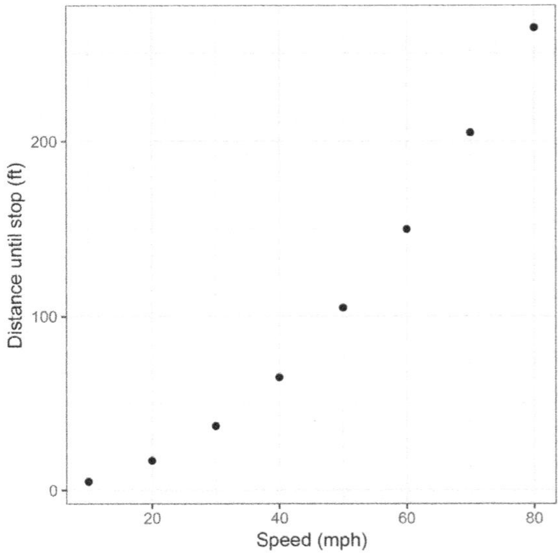

Possible answer: *It appears there is a moderate to strong positive linear relationship between speed and the distance until stopped, although the scatterplot appears to have a slight curve. The actual value for r = 0.97.*

Further Explorations and Extensions

The formula for calculating Pearson's correlation coefficient is:

$$r = \frac{1}{n-1}\sum_{i=1}^{n}\left(\frac{X_i - \bar{X}}{s_X}\right)\left(\frac{Y_i - \bar{Y}}{s_Y}\right)$$

In the formula, X_i and Y_i are the individual data points, $\bar{X}$ and $\bar{Y}$ are the means of the explanatory values and response values, and s_x and s_Y are the standard deviations of the explanatory values and response values. This formula is not necessary for students to memorize or calculate by hand.

The following explanation can be used to develop the conceptual understanding of this formula. Using technology, draw a horizontal line through the mean of the earnings ratio and draw a vertical line through the mean of the years since 1970.

Number each of the four quadrants, I (upper right), II (upper left), III (lower left), and IV (lower right).

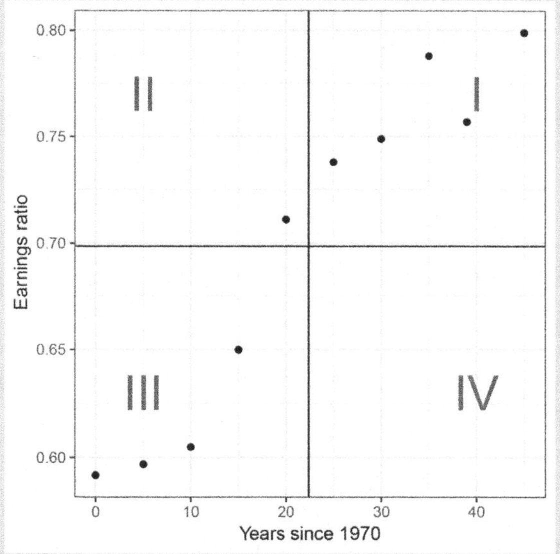

Ask your students what they observe in terms of the number of points in each region.

Possible response: *Almost all the points are in regions I and III.*

Further Explorations and Extensions Cont.

Have students focus on the points in Region I. Ask what is true about the points in region I as compared with the point ($\bar{X}$ and $\bar{Y}$).

Answer: *These points have x values greater than the mean of the incomes for men and y values greater than the mean of the incomes for women.*

Point to the correlation coefficient formula. Ask how the ordered pairs in Region I result in a positive component in the sum part of the formula.

Answer: *The $(X_i - \bar{X})$ will be positive and $(Y_i - \bar{Y})$ will be positive so the product will be positive.*

Ask students what will happen in the formula for the points in Region III.

Answer: *The ordered pairs in Region III result in a positive component in the sum part of the formula. The $(X_i - \bar{X})$ will be negative and the $(Y_i - \bar{Y})$ will be negative, so the product will be positive.*

Ask students what will happen in the formula for the points in Region IV.

Answer: *The ordered pairs in Region IV result in a negative component in the sum part of the formula. The $(X_i - \bar{X})$ will be positive and $(Y_i - \bar{Y})$ will be negative, so the product will be negative.*

Ask students what will happen in the formula for the points in Region II.

Answer: *The $(X_i - \bar{X})$ will be negative and $(Y_i - \bar{Y})$ will be positive, so the product will be negative.*

Ask what the sign of the sum of all the products will be. How does this relate to the graph? Explain.

Answer: *The sign of the sum of all the products will be positive because there are only positive values since all points but one lie in quadrants I and III.*

Ask students when a correlation coefficient would be close to zero.

A correlation close to zero occurs when there are offsetting positive and negative products, resulting in a sum close to zero and no apparent trend in the data.

Investigation 7

Are Gender and Pay Related? Continued

Assessing Linear Fit

Overview

This investigation continues to explore relationships within bivariate data, using the same data set as in Investigation 6, which explored the relationship between median income and gender.

The previous investigation focused on using the correlation coefficient to determine the strength of the linear relationship between two quantitative variables (earnings ratio vs. years since 1970). This investigation builds on the idea of error developed in Investigation 5 (the amount of error between the actual and guessed number of calories in a small candy bar) and furthers this concept by defining residuals and exploring the use of residual plots as a tool to determine the appropriateness of using a line of fit for bivariate data.

This investigation follows the four components of statistical problem solving put forth in the *Guidelines for Assessment and Instruction in Statistics Education (GAISE) Report*. The four components are formulate a question, design and implement a plan to collect data, analyze the data by measures and graphs, and interpret the results in the context of the original question. This is a GAISE Level B activity.

The Analyze the Data section of the student worksheet is divided into three parts. The first part is optional, depending on the background knowledge of the students. It allows students to fit a line to data, determine the equation of the line, interpret the slope, and introduce the concept of error. The second part of this section uses statistical software to demonstrate the development of the concept of a residual. The third part continues to use statistical software to demonstrate how to construct residual plots.

Instructional Plan

Brief Overview

» Continue with the statistical question: "To what extent are the median income of males and the median income of females related?"

» Create linear models to fit a data set and explore the least squares regression line.

» Define residuals and create and use a residual plot to determine whether a linear model is appropriate for a data set.

» Answer the statistical question.

Part I: Fitting a Line to Data

Scenario

The scenario is the same as was explored in Investigation 6, which should be completed before this investigation.

Formulate a Statistical Question

(from Investigation 6)

Continue with the statistical question: "To what extent are the median income of males and the median income of females related?"

Collection of Data

(from Investigation 6)

Learning Goals

- Interpret the slope (rate of change) and intercept (constant term) of a linear model in the context of the data.
- Represent data on two quantitative variables on a scatterplot and describe how the variables are related.
- Fit a linear function for a scatterplot that suggests a linear association.
- Informally assess the fit of a function by plotting and analyzing residuals.

Mathematical Practices Through a Statistical Lens

MP7. Look for and make use of structure.

Students use structure to separate the 'signal' from the 'noise' in a set of data—the 'signal' being the structure, the 'noise' being the variability. They look for patterns in the variability around the structure and recognize these patterns can often be quantified.

Materials

Student worksheets are available at *www.statisticsteacher.org/statistics-teacher-publications/focus.*

Part I

- Straightedge (ruler or piece of spaghetti)
- Student Worksheet 7.1 Fitting a Line to Data (Used for Part I of Analyze the Data section)

Parts II and III

- Statistical technology that can do the following:
  - Create a scatterplot
  - Create horizontal and vertical lines at defined points
  - Create a moveable line
  - Show the residuals, squares of the residuals, and sum of the squares of the residuals
  - Create the least squares regression line

(Core Math Tools is one piece of statistical software that can be used for demonstration: www.nctm.org/Classroom-Resources/Core-Math-Tools/Core-Math-Tools or http://nctm.org/resources/cmt/CoreMathTools.jar)

- Exit Ticket

Estimated Time

Two 50-minute class periods for parts I, II, and III. Part I is optional, depending on background knowledge of students.

One 50-minute class period for parts II and III. If only doing parts II and III, Part II may take a little longer (possibly 1.5 class periods total), depending on the comfort level of students with the technology being used.

Pre-Knowledge

Students should be able to:

» Construct scatterplots

» Given two ordered pairs, write the equation of a line in slope-intercept form

» Interpret the slope of a line in context of the given data

Table 7.1

Years since 1970	Median Income Men ($)	Median Income Women ($)	Earnings ratio
0	9,184	5,440	0.592
5	12,934	7,719	0.597
10	19,173	11,591	0.605
15	24,999	16,252	0.650
20	28,979	20,591	0.711
25	32,199	23,777	0.738
30	38,891	29,123	0.749
35	42,188	33,256	0.788
39	49,164	37,234	0.757
45	50,119	40,022	0.799

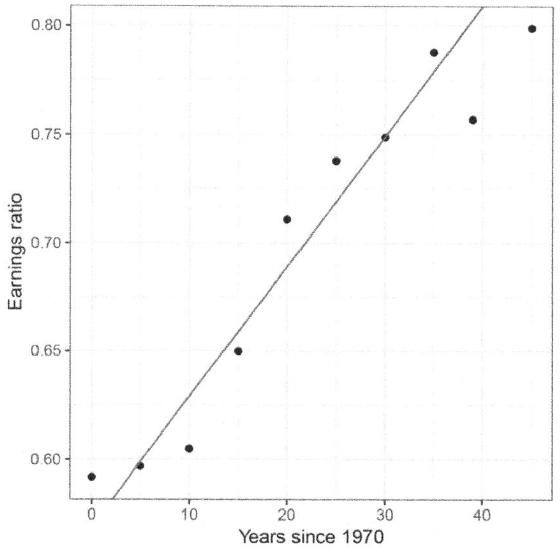

Figure 7.1: Scatterplot with example of line drawn

Analyze the Data

Part I: Informally Fitting a Line to Data

Note: If students have had experiences fitting a line to data, writing the equation of the line, and interpreting the slope of the line in context, this part can be skipped.

Distribute Student Worksheet 7.1 Fitting a Line to Data and a straightedge (piece of spaghetti or ruler).

Ask your students to complete numbers 1 to 5.

Example: Figure 7.1

1. Using a straightedge, draw a line on the scatterplot you think will summarize or fit the data. Explain how you decided to draw your line.

Possible answers:

» *Balanced same number of points above and*

below the line (misconception)

» *Connected a point in the bottom left corner to a point in the upper right corner*

» *Found the mean of the x's and y's to create a midpoint first, then fit a line through that point (unlikely but a great strategy)*

2. Locate two points on the line. Write each point as an ordered pair.

3. Using the two points, write the equation of your line in slope-intercept form.

Possible answer: $\hat{y} = 0.006x + 0.569$

4. Interpret the slope of your line in terms of the context.

Possible answer: *The slope means that for each additional year, the earnings ratio is predicted to increase about 0.006, on average. The earnings ratio is the median women's income to the median men's income, which means the women's income is increasing by $0.006 for every $1 of men's income each year.*

5. Interpret the y-intercept in terms of the context.

Possible answer: *The y-intercept of 0.569 would mean the predicted earnings ratio is 0.569 in 1970.*

Discuss the answers for questions 1 to 5.

Ask students to complete questions 6 to 10 on Student Worksheet 7.1.

6. Using 25 years from 1970 as your x value and the equation of your line, what is your prediction for the earnings ratio?

Possible answer: $0.006(25) + 0.569 = 0.719$

7. Look at the data table to find the actual earnings ratio of 25 years from 1970.

Answer: *0.738*

8. What is the difference between the actual earnings ratio for 25 years from 1970 and the earnings ratio your line predicted?

Possible answer: $0.738 - 0.719 = 0.019$

9. How well did your line predict the earnings ratio for 25 years from 1970? Did your line overestimate or underestimate the earnings ratio?

Possible answer: *close estimate but slightly underestimated*

10. Compare your prediction with others in class. Who was the best predictor of the earnings ratio for 25 years since 1970?

Possible answer would be to use the same strategy used in Investigation 5 to find the smallest error of the guesses.

Discuss answers to questions 6 to 10 with the whole class.

Note: If students need more experiences fitting a line to data, see Section 5 of *Bridging the Gap Between Common Core State Standards and Teaching Statistics* by Hopfensperger, Jacobbe, Lurie, and Moreno, published by the American Statistical Association in 2012 and available at *ww2.amstat.org/education/btg/index.cfm.*

Part II: Fitting a Line to Data Using Technology

Note: Part II is best done as a demonstration on a screen with the whole class while individual students or pairs of students have access to the same technology on a device. One suggestion is to have a student lead the class on the technology while the teacher gives supporting directions.

A free statistical application that can be used for the demonstration is Core Math Tools, available at *www.nctm.org/Classroom-Resources/Core-Math-Tools/Core-Math-Tools* or *http://nctm.org/*

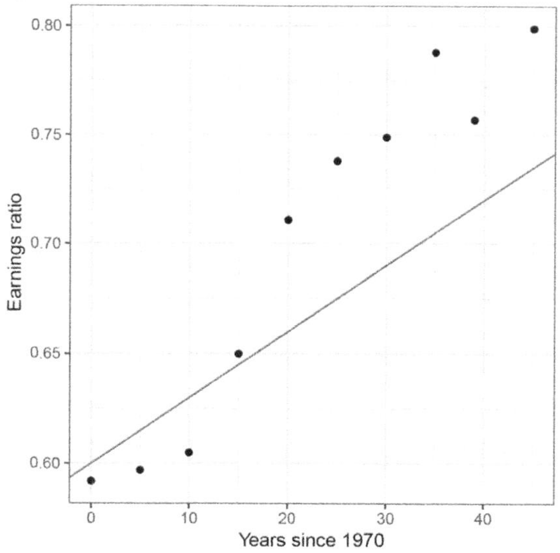

Figure 7.2: Scatterplot of earnings ratio vs. years since 1970 with a moveable line

resources/cmt/CoreMathTools.jar. However, any statistical technology can be used, as long as it meets the requirements in the materials list.

Other free tools:

» *www.rossmanchance.com/applets/RegShuffle.htm*

» *www.stapplet.com*

» *www.lock5stat.com/StatKey*

Before the demonstration, the data (years since 1970 and corresponding earnings ratio) need to be entered into the statistical software being used for the demonstration with years since 1970 as the independent and earnings ratio as the dependent variable.

Display the scatterplot of earnings ratio vs. years since 1970.

Remind your students that they determined there was a strong positive correlation (r=0.963) and the data appear to fit a line in Investigation 6.

Explain that the next step in the analysis of the data is to find the equation for a line that best fits the data.

Explain that we first want to draw an approximate line of fit using technology. To begin, we are going to "eyeball" a line.

Demonstrate how to add a moveable line to the scatterplot (See Figure 7.2).

After the line is on the graph, point out the equation of the line. (In this example, the equation is $\hat{y} = 0.006x + 0.569$). Move the line around, changing both the position and the slope.

Note: Usually, the line can be moved by dragging the small square near the center of the line (changes the intercept) and the small squares toward the ends of the line (changes the slope). The equation of the line is in the upper right corner of the screen.

Ask students to direct which way(s) to move the line to fit the data.

Once students have decided the line is in the "best" place, have them interpret the slope and y-intercept of their line in context.

Possible answer: *For the example above, the slope means the earnings ratio is predicted to increase about 0.006, on average, for each additional year. The earnings ratio is the median women's income to the median men's income, which means the women's income is increasing by \$0.006 for every \$1 of men's income each year. The y-intercept of 0.569 would mean the predicted earnings ratio is 0.569 in 1970.*

Point out that this equation of the line of fit can be used to make predictions.

Write the equation of the line on the board using the symbol $\hat{y}$. For example, $\hat{y} = 0.006x+0.569$. The symbol $\hat{y}$ (y-hat) stands for the *predicted* value for y for a given value of x.

Ask your students to use the equation of the line and predict the value of the earnings ratio for 25 years since 1970.

Possible answer: *0.006(25) + 0.569 = 0.719*

Ask your students how well their line predicts the earnings ratio for 25 years from 1970. Did their line overestimate or underestimate the earnings ratio?

Possible answer: *Close estimate but slightly underestimated*

Ask your students if they think they have found the "best" line. Ask them how they could decide if they have the "best" line.

Ask your students how they determined who was the "best guesser" when guessing the number of calories in candy bars (Investigation 5)?

Possible answer: *Found the difference between actual calories and guessed number of calories, squared the differences, and found the sum of the differences. The lowest sum of the squared differences was determined to be the "best" guesser.*

Explain that this same strategy will be used to determine the line of "best" fit.

Introduce the term "residual" by sharing the definition and a drawing.

A *residual* is the vertical (signed) distance between a data point and the graph of the regression equation. The residual is positive if the data point is above the graph. The residual is negative if the data point is below the graph. The residual is 0 only when the graph passes through the data point.

Notation: (x_i, y_i) is used to represent any data point where i represents the ith data point. For example, (x_2, y_2) is the second data point. The

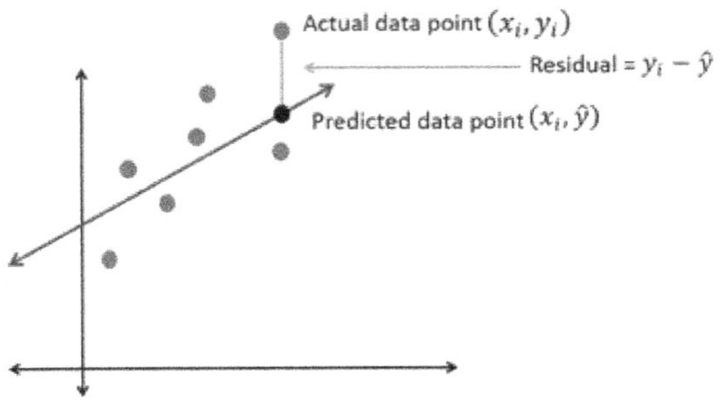

Figure 7.3: A drawing of a residual

notation of $\hat{y}$ (read as "y-hat") represents the predicted y-value from the graph of the regression line. Thus, the residual is $y_i - \hat{y}_i$, or *observed value – predicted value* (See Figure 7.3).

Explain that the line of best fit, referred to as the *least squares regression line,* is defined as the line that has the lowest sum of the squared residuals, $\Sigma\,(y_i - \hat{y}_i)^2$.

Use the software and display all the residuals on the moveable line. As the line moves, the residuals should change (See Figure 7.4).

Show the squares that are formed by the residuals (See Figure 7.5).

Remind your students they determined the best guesser of calories by finding the sum of

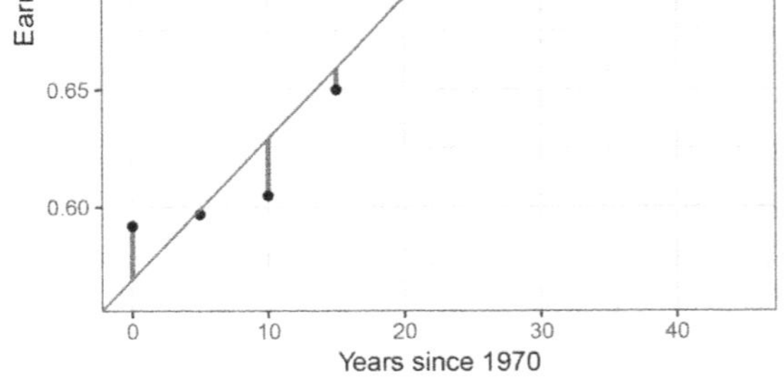

Figure 7.4: The residuals on the moveable line

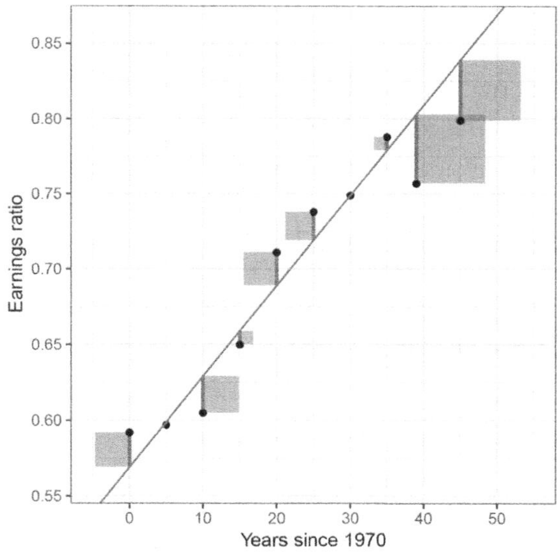

Figure 7.5: The squares formed by the residuals

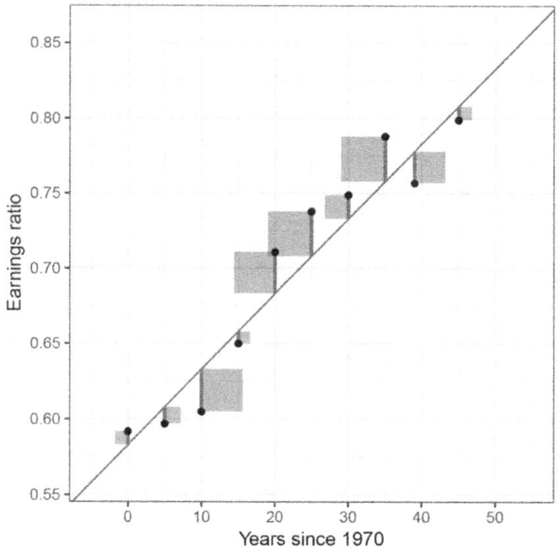

Figure 7.6: The least squares regression line

the squared differences between their guesses and the actual number of calories. Explain that, in a similar fashion, we want to find the sum of the squared residuals or the sum of the areas of the squares shown on the graph to find the "best" line.

Use the software to calculate and display the sum of the squared residuals.

Demonstrate that, as the line moves, the squares change and the sum of the residuals squared changes. Continue to move the line until students think the lowest sum of residuals squared has been found. If students are working on technology, allow them time to explore and find what they think is the lowest sum of squared residuals.

Explain that the software will actually find the line with the lowest sum of squared residuals.

Remove the moveable line. Display the least squares regression line. Show the squares and point out the sum of the squared residuals. Ask how well the class did in finding the least squares line using the moveable line feature of the software (See Figure 7.6).

Point out the equation $\hat{y} = 0.005x + 0.583$ or *predicted earnings ratio* = 0.583+0.005(*years since* 1970) is the *least squares regression line.* This is the line that minimizes the sum of the squared residuals. The symbol $\hat{y}$(y-hat) stands for the predicted value for y for a given value of x.

Part III: Residual Plots

When investigating the relationship between two quantitative variables, the first step was to construct a scatterplot and look for any relationship between the variables. If it appeared linear was an appropriate model, the correlation coefficient was used to determine the strength of a linear association. To get a closer look at the deviations from the line that may not be seen in the scatterplot, the next step is to construct a residual plot.

A ***residual plot*** is a graph that shows the residuals on the vertical axis and the explanatory variable on the horizontal axis. If the points in a residual plot are randomly dispersed around the horizontal line through the residual = 0 or $\hat{y} = 0$, a linear regression

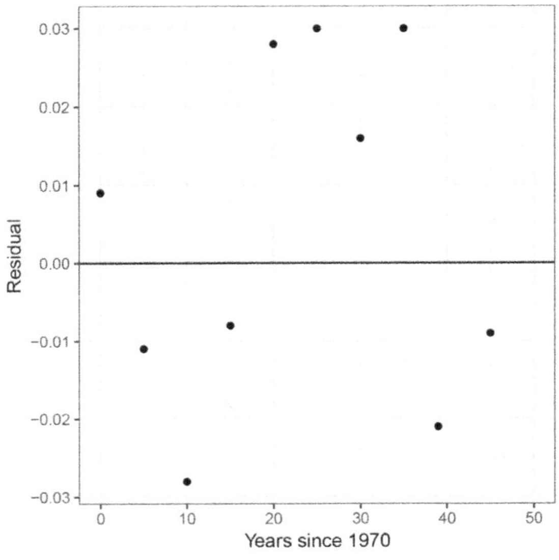

Figure 7.7: A residual plot

model is appropriate for the data; otherwise, a nonlinear model is more appropriate.

Create a residual plot using the statistical software (See Figure 7.7).

If a linear model is a good fit for the data, then the residual plot should show a random dispersion around the line, residual = 0, or $y_i - \hat{y}_i = 0$.

For this set of data, the residual plot shows no pattern in the residuals, further indicating a linear model is appropriate for the data.

Choose a point on the residual plot and ask students to explain what this point represents and how it is connected to the data set and least squares regression line. It might be useful to show the residual plot next to the graph that shows the residuals on the least squares regression line and point out where each residual can be seen on each graph (See Figure 7.8). Ask students to explain what the graph represents.

Interpret the Results in the Context of the Original Question

Ask your students to restate the statistical question. To what extent are the median income of males and median income of females related?

Ask your students to complete problems 11 and 12.

11. Talk with a partner and then write a paragraph to answer the statistical question based on the analysis that refers to the residuals.

Optional: What might be some social implications of the results?

Students might just add to the paragraphs they wrote in Investigation 6.

Possible answer: *A scatterplot of the data shows a positive linear pattern. The correlation coefficient is approximately 0.963, indicating a strong positive correlation between the years since 1970 and earnings ratio of median in-*

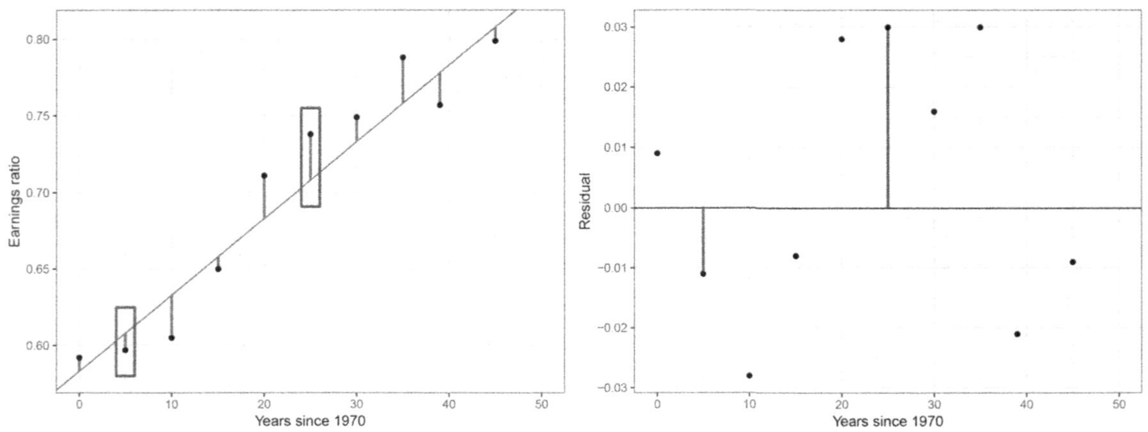

Figure 7.8: Residual plot next to the graph showing the residuals on the least squares regression line

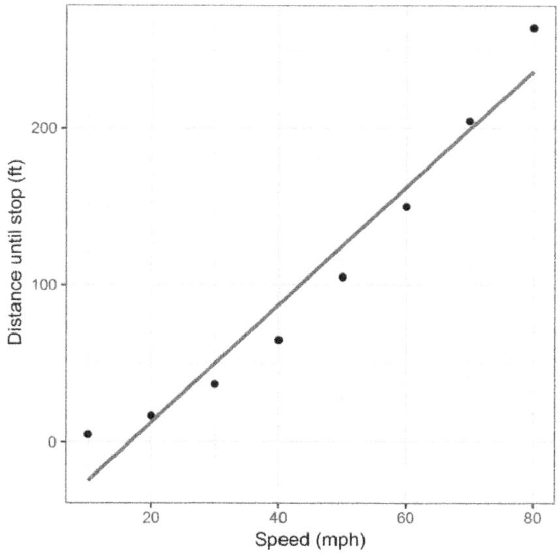

Figure 7.9: A scatterplot showing when linear would not be a good model

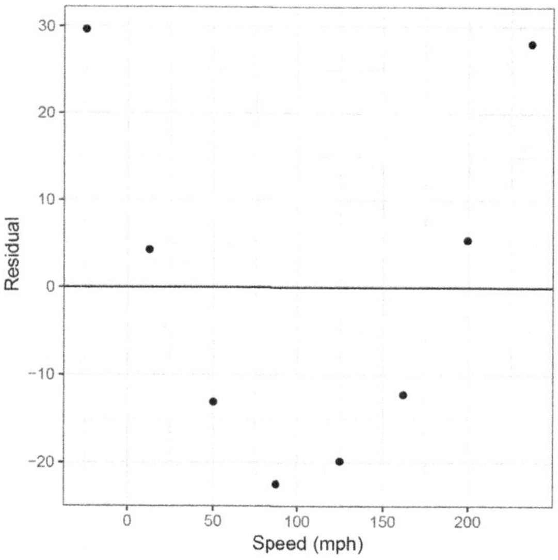

Figure 7.10: A residual plot showing when linear would not be a good model

comes of women to men. The equation for the line of best fit is (predicted earnings ratio) = 0.583+0.005(years since 1970), indicating the earnings ratio increases approximately .005 each year since 1970. The earnings ratio was 0.583 in 1970, according to the linear model. The residual plot shows no pattern, indicating a linear model is appropriate. If the pattern were allowed to continue, the earnings ratio will eventually be 1, indicating no gender gap in income.

12. When is the earnings ratio predicted to be 1? Do you think this will happen?

Ask students to find out when the earnings ratio would be 1 (when men and women have the same median income), as predicted by the line of best fit.

Possible answer: *The line of best fit predicts the earnings ratio will be 1 between 2053 and 2054. Answers will vary about whether this will happen. There are many variables that could affect the earnings ratio in the next 20+ years.*

Note: It might be useful to show a residual plot in which a pattern does occur. The scatterplot (Figure 7.9) and residual plot (Figure 7.10) show an example in which a pattern occurs in the residual plot and linear would not be a good model.

Example: Braking Road Test

Road tests under various conditions are likely to produce data like these that show speed (in mph) and distance until stopping (in feet).

While it appears a linear model might be appropriate from the graph of the data, the residual plot shows a pattern, indicating a linear model is not appropriate. The correlation coefficient can be calculated, and $r \approx 0.976$. This indicates linearity is plausible. Without looking at a residual plot, it might be difficult to tell a linear model is not appropriate.

Additional Ideas

Note that this lesson could focus on a different source of data, such as the cost of a 30-second Super Bowl ad and the winning team player's share (data set available through Core Math Tools).

Exit Ticket

For each scatterplot shown in problems 1 and 2:

a. Sketch a residual plot from the line of best fit
b. Determine if a linear model is appropriate
c. If a linear model is appropriate, estimate the correlation coefficient
d. Summarize the relationship between the variables
e. Interpret the values in the equation of the line of best fit in the context of the situation

Extra: What statistical question might the data answer?

1. **Health and Nutrition**
   The scatterplot in Figure 7.11 shows how average daily food supply (in calories) is related to life expectancy (in years) in a sample of countries in the western hemisphere. The equation of the least squares regression line is $\hat{y} = 0.009x + 4681$.
   Source: World Health Organization Global Health Observatory Data Repository, faostat3.fao.org

a. Sketch a residual plot from the line of best fit

Answer: *Figure 7.12*

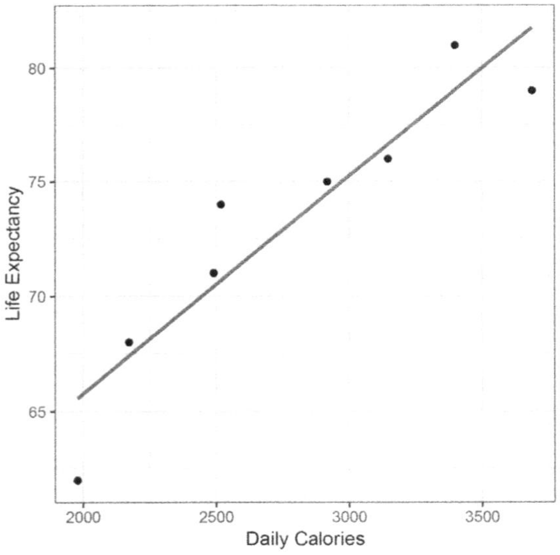

Figure 7.11: Average daily food supply (in calories) related to life expectancy (in years) in a sample of countries in the western hemisphere.

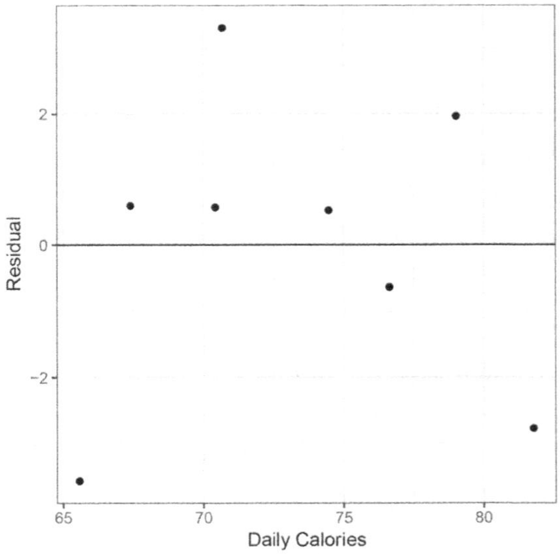

Figure 7.12: Sketch of a residual plot from the line of best fit

Exit Ticket Cont.

b. Is linear an appropriate model? Explain

Possible answer: *The residual plot is fairly scattered and shows no pattern, indicating a linear model is an appropriate model.*

c. Estimate for correlation coefficient

Possible answer: *The correlation coefficient could be estimated at approximately 0.8-0.9, ($r \approx 0.928$).*

d. Summary of the relationship

Possible answer: *The relationship indicated by the graph is a positive linear relationship, indicating the average life expectancy increases as daily caloric intake increases.*

e. Interpretation of slope and intercept

Possible answer: *The slope of 0.009 means the life expectancy increases 0.009 years for every 1 calorie, or 0.9 (almost 1) years of life expectancy is gained for every increase of 100 calories of daily intake. The y-intercept of about 47 means the life expectancy would be 47 years for someone with a daily caloric intake of 0 calories. In this case, the y-intercept does not make much sense in the context, as a person intaking 0 calories daily would probably not live for 47 years.*

Extra: A statistical question might be: "To what extent are daily caloric intake and life expectancy related?"

2. Crawling Age

The scatterplot in Figure 7.13 shows the average daily outside temperature when the babies were six months old, and the average age in weeks at which those babies began to crawl are reported. The equation of the least squares regression line is $\hat{y} = -0.078x + 35.68$.

(Source: Benson, Janette. "Infant Behavior and Development," 1993.)

a. Sketch a residual plot from the line of best fit

Answer: *Figure 7.14*

b. Is linear an appropriate model? Explain

Exit Ticket Cont.

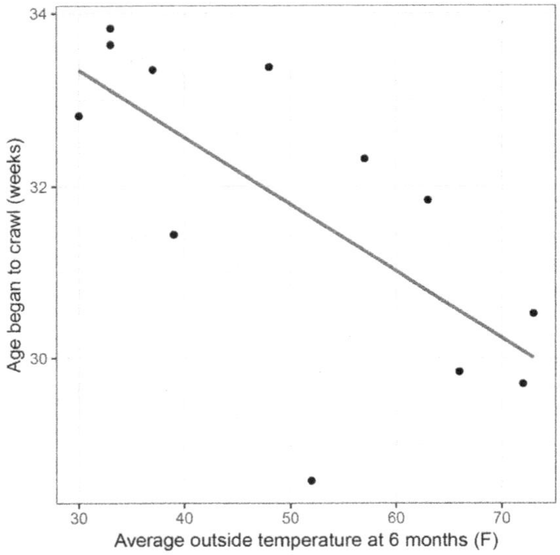

Figure 7.13: Average daily outside temperature when babies were six months old and average age in weeks at which those babies began to crawl

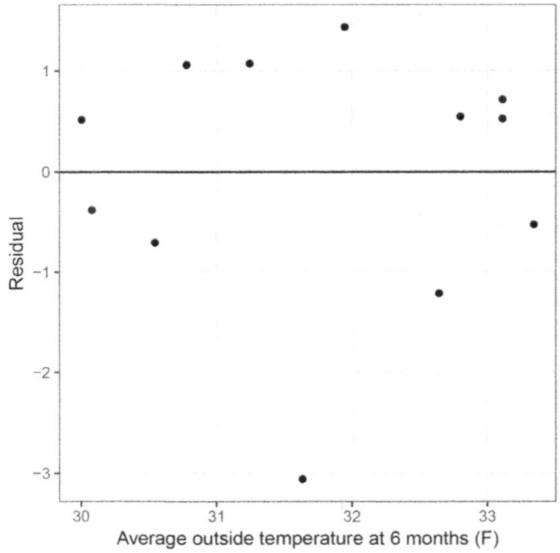

Figure 7.14: Sketch of a residual plot from the line of best fit

Possible answer: *The residual plot is fairly scattered and shows no pattern, indicating a linear model is an appropriate model.*

c. Estimate for correlation coefficient

Possible answer: *The correlation coefficient could be estimated between -0.6 and -0.8, ($r \approx -0.7$).*

d. Summary of the relationship

Possible answer: *The relationship indicated by the graph is a negative linear relationship, indicating the average age at which babies begin to crawl decreases as the average outside temperature increases.*

e. Interpretation of slope and intercept

Possible answer: *The slope of -0.078 means the average age at which a baby begins to crawl decreases by 0.078 weeks for every increase of 1 degree outside temperature. The y-intercept of about 36 means the average crawling age would be 36 weeks when the average outside temperature is 0 degrees. There are probably not many places where the average outside temperature is 0 degrees, but that might occur in locations closer to the north or south poles.*

Exit Ticket Cont.

Extra: A statistical question might be: "To what extent are average outside temperature and average crawling age of babies related?"

Note: Point out to students that even though there is a negative correlation between average outside temperature and crawling age, high outside temperatures do not necessarily cause the crawling age to decrease.

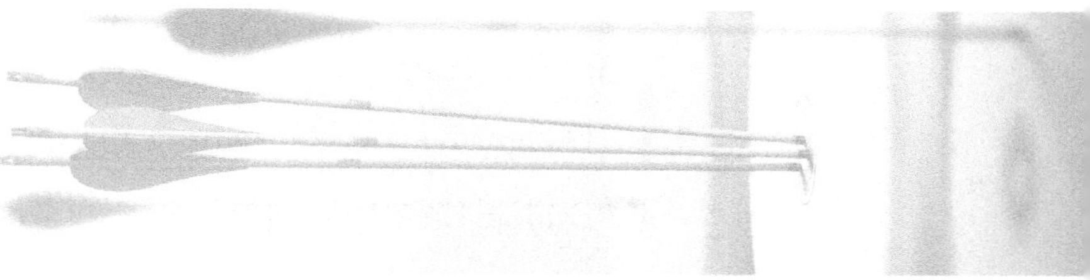

Investigation 8

How Long to Topple Dominoes?

Exploratory Lesson

Overview

This investigation offers two options for students. The first provides students an opportunity to use the four components of statistical problem solving by designing their own investigation around a topic of interest that involves exploring a relationship between two quantitative variables. Several suggestions are included in this investigation, and students could be encouraged to come up with their own variables.

Encourage students to work in pairs or small groups. The results could include written and oral presentations and/or construction of a poster to display the data and answer the statistical question. Information about creating a statistical poster, rubric, and competition information can be found at *www.amstat.org/asa/education/ASA-Statistics-Poster-Competition-for-Grades-K-12.aspx.*

The second option provides the set of directions for an investigation titled "How Long to Topple Dominoes?" This option is suggested for students who may need more scaffolding and direction when collecting and analyzing data. This option is based on Lesson 11, Exploring Linear Relations, available at *www.amstat.org/asa/files/pdfs/ddmseries/ExploringLinearRelations.pdf.*

Both options for this investigation follow the four components of statistical problem solving put forth in the *Guidelines for Assessment and Instruction in Statistics Education (GAISE) Report.* The four components are formulate a question, design and implement a plan to collect data, analyze the data by measures and graphs, and interpret the results in the context of the original question. This is a GAISE Level B or C activity, depending on the amount of scaffolding provided.

Instructional Plan

» This lesson can be open-ended and allow students to choose a topic, as long as it involves two quantitative variables anticipated to have a linear relationship.

» Students follow the statistical problem-solving process, guided by the teacher. Provide support when needed.

» Students could be required to create a poster, a presentation, and/or a written report to communicate their process and results.

Instructions for Design Your Own Investigation

Explain that students will follow the four steps of the statistical problem-solving process. Have students work in pairs or small groups. Distribute Student Worksheet 8.1 Design Your Own Investigation.

Formulate a Statistical Question

Students will brainstorm topics that might interest their group and include two variables. The two variables should be quantitative and ones in which students anticipate a linear relationship. Possible ideas include the following:

» Describe the relationship between two body measurements. Possible measurements

Learning Goals

- Design and collect data from an experiment
- Explore the relationship between two quantitative variables
- Apply techniques of finding a linear model
- Analyze the fit of the linear model

Mathematical Practices Through a Statistical Lens

MP1. Make sense of problems and persevere in solving them.

Statistically proficient students understand how to carry out the four steps of the statistical problem-solving process: formulating a statistical question, designing a plan for collecting data and carrying out that plan, analyzing the data, and interpreting the results.

Materials

Student worksheets are available at *www.statisticsteacher.org/statistics-teacher-publications/focus.*

Option One

- Student Worksheet 8.1 Design Your Own Investigation

Option Two

- Student Worksheet 8.2 Topple Dominoes
- 200 dominoes (about 30 for each student group)
- Meter stick for each group
- Stopwatch for each group
- YouTube video of dominoes toppling: *www.youtube.com/watch?v=y4VJssQv_Qw*

Estimated Time

One to three 50-minute class periods, depending on the final product, amount of work required outside of class, and whether presentations occur.

Pre-Knowledge

Students should be able to:

- Find and interpret the equation of the least squares regression line
- Find and interpret the correlation coefficient
- Construct and interpret a residual plot

include width of head, width of shoulders, length of forearm (elbow to wrist), length of upper arm (elbow to shoulder), top of head to navel, top of head to tip of fingers (arms at sides), wrist circumference, neck circumference, height, arm span (arms out to sides, tip of fingers to tip of fingers), foot length, and stride length.

» Describe the relationship between ramp height and distance a matchbox car travels.

» Describe the relationship between number of people and length of time to pass a stack of books, pass a bucket of water, or bounce and pass a ball.

» Describe the relationship between length of time to say a tongue twister and the number of people.

» Describe the relationship between height of a catapult and how far a gummy bear is launched (Lesson 12, Exploring Linear Relations, *www.amstat.org/asa/files/pdfs/ddmseries/ExploringLinearRelations.pdf*).

Students should then develop the statistical question. Students could check in for approval before moving on to collect data.

Collect Appropriate Data

Have students describe the data-collection process, including possible complications and how these might be handled. Students should check in for approval before collecting data. Then, have students collect and organize the data.

Analyze the Data

Data analysis should include a scatterplot, description of the relationship between the two variables, interpretation of correlation coefficient, linear model, and residual plot.

Interpret the Results in the Context of the Original Question

Interpret the analysis of the data in the context of the situation. Be sure to answer the statistical question and support the answer with the data analysis.

Option 1: Write and orally present a report summarizing your results. Your report and presentation should include the following:

» The statistical question investigated and why it was chosen

» A description of the population sampled

» A summary of the data collection

» The collected data, organized as appropriate

» Analysis and descriptions of the data, using calculations, tables, graphs, and plots. Note any unusual results.

» Conclusions about the statistical question

» Recommendations for any follow-up studies or questions that may be investigated

Option 2: Create a data visualization poster and orally present the poster summarizing your results.

The poster should include the following:

» The statistical question as the title of the poster

» The organized collected data—tables and graphs (at least two graphs)

» Conclusions about the statistical question

The oral report should include the following:

» Reason the statistical question was chosen

» A description of the population sampled

Data Collection Table

Number of Dominoes	Time (Sec)	Time (sec)	Time (sec)	Mean Time (sec)
10				
15				
20				
25				
30				

- » A summary of the data collection
- » Analysis and descriptions of the data using calculations, tables, graphs, and plots. Note any unusual results.
- » Recommendations for any follow-up studies or questions that may be investigated.

Instructions for How Long to Topple Dominoes?

Scenario

On November 13, 2009, World Domino Day 2009 saw the world record broken for the most dominoes toppled by a group when 4,491,863 dominoes were toppled. A total of 89 builders set up the dominoes in the WTC Expo Center in Leeuwarden, The Netherlands.

In April of 2017, a group of three students broke the unofficial world record for longest domino line with 15,524 dominoes!

View a YouTube video of the dominoes toppling at *www.youtube.com/watch?v=y4VJssQv_Qw.*

Formulate a Statistical Question

How long do you think it took for the line of 15,524 dominoes to fall over?

How long do you think it took for the 4,491,863 dominoes to fall over?

Statistical question: "What is the relationship between the number of dominoes in a line and the length of time for all the dominoes to topple over?"

Collect Appropriate Data

Divide students into groups of three or four. Distribute Student Worksheet 8.2 Topple Dominoes and 30 or more dominoes to each group. Each group should also have a meter stick and stopwatch. Students will need a flat and hard surface to set up dominoes. Carpeting does not work well.

Directions

On a flat and hard surface, stand up the dominoes on end in a straight line. Use the ruler or meter stick to make the spacing between the dominoes even. Space the dominoes about 2.5 cm apart. The only rule is that a domino can knock over only one other domino when it falls.

Set up 10 dominoes in a straight line and carefully time how long it takes for all 10 dominoes to topple. Repeat two more times. Record the three times in the data collection table.

Set up 15 dominoes in a straight line and carefully time how long it takes for all 15 dominoes to topple. Repeat two more times. Record the three times in the data collection table.

Set up 20 dominoes in a straight line and carefully time how long it takes for all 20 dominoes to topple. Repeat two more times. Record the three times in the data collection table.

Set up 25 dominoes in a straight line and carefully time how long it takes for all 25 dominoes to topple. Repeat two more times. Record the three times in the data collection table.

Set up 30 dominoes in a straight line and carefully time how long it takes for all 30 dominoes to topple. Repeat two more times. Record the three times in the data collection table.

Analyze the Data

1. Using technology construct a scatterplot of the mean time for the dominoes to fall versus the number of dominoes. Make a sketch of the scatterplot.
2. Is a linear model appropriate to describe the relationship between time for all the dominoes to fall and the number of dominoes? Use the correlation coefficient and a residual plot to explain your reasoning.
3. Find the equation of the least squares regression line.
4. Interpret the slope of the least squares regression line in context.

Interpret the Results in the Context of the Original Question

5. Using the linear model you developed, make a prediction for how long it would take 15,524 dominoes to fall. To make this prediction, what assumptions do you have to make about your model?
6. Watch the video at *www.youtube.com/watch?v=y4VJssQv_Qw* again and time how long it takes for all the dominoes to fall over.
7. How close was your prediction? What are some reasons why your prediction might have been off?

Extensions

Calculate the speed the dominoes travel as they topple.

Design and conduct an experiment to investigate the relationship between the distance between the dominoes and effect on time.

Investigation 9

Survey Says?

Analyzing Categorical Data in a Statistical Study

Overview

This investigation is the first of several lessons that focus on the analysis of two categorical variables. This investigation shares an initiative carried out by students at an urban high school that involved collecting data from the student body. This initiative involved creating a survey, obtaining a sample of completed surveys, and analyzing the sample to answer statistical questions posed by the students. The students addressed whether the sample was representative of the student body of their school.

The four components of statistical problem solving as put forth in the *Guidelines for Assessment and Instruction in Statistics Education (GAISE) Report* are addressed in this investigation. The four components are formulate a statistical question, design and implement a plan to collect data, analyze the data by measures and graphs, and interpret the results in the context of the original question. This investigation is a GAISE Level B activity.

Instructional Plan

Brief Overview

» Define and give examples of categorical data.

» Read the scenario and study the survey questions.

» Develop a statistical question based on the survey questions.

» Discuss the four options to collect survey data.

» Discuss the collection plan used by high-school students.

» Summarize survey results.

Introduction to Categorical Data

Ask your students to give examples of *numerical* data they have analyzed. Have them share what types of graphs and calculations they did in their analysis of numerical data.

Possible answer: *Examples from previous investigations include height and arm span, length of baseball games in hours, time to complete a memory game, and homework times. Students would have constructed box plots, dot plots, and histograms and found means, medians, IQRs, and standard deviations.*

Discuss another type of data called *categorical data*. Categorical variables take on values that are names or labels. Share some examples of categorical data such as an answer to a true or false question, an answer to a multiple-choice question (A, B, C, or D), the size of a T-shirt (small, medium, or large), or the breed of a dog (shepherd, terrier, lab).

Ask your students to share other examples of categorical data.

Hand out Student Worksheet 9.1 Scenario

Explain that the case study presented was conducted by high-school students from an urban high school who analyzed categorical data.

Direct students to read the scenario from Student Worksheet 9.1.

Scenario

The administration at Rufus King High School, a United States urban high school of students in grades 9 to 12, was in the process of evaluating the school's academic and extracurricular programs. The high-school administration considered distributing and analyzing a survey addressing the school's programs that would be similar to the process businesses use to evaluate their products and services. They asked the students enrolled in an 11th grade mathematics class if they would help with the design, distribution, and analysis of a survey project.

Statistical studies about a school's services might result in decisions that alter a school's daily schedule, curriculum, course offerings, extracurricular opportunities, etc. Rufus King students wanted to be part of a study that might alter their school's academic and extracurricular programs. Students designed

Learning Goals

» Investigate methods for obtaining a representative sample of responses to a survey from a large population.

» Evaluate a random sample of students' responses to a survey.

» Summarize a population using the results of a random sample.

Mathematical Practices Through a Statistical Lens

MP2. Reason abstractly and quantitatively.

Statistically proficient students are able to summarize data to answer statistical questions. Students explain their summaries of data using proportions.

Materials

Student worksheets are available at *www.statisticsteacher.org/statistics-teacher-publications/focus*

» Student Worksheet 9.1. Scenario

» Student Worksheet 9.2 Data Collection Methods

» Student Worksheet 9.3 Survey Results

» Exit Ticket

Estimated Time

One 50-minute class period. This lesson introduces a case study that is expanded in investigations 10 and 11.

Pre-Knowledge

Students should be able to convert a proportion to a percent.

Question 1:	Indicate your gender:
	___ Female (F) ___ Male (M) ___ Prefer not want to respond
Question 2:	Indicate your grade level in high school:
	___ 9th grade ___ 10th grade ___ 11th grade ___ 12th grade
Question 3:	Do you consider yourself a dog person, a cat person, or neither?
	___ A. I consider myself a dog person.
	___ B. I consider myself a cat person.
	___ C. I do not consider myself a dog or cat person.
Question 4:	What is your main goal after completing high school?
	___ A. To attend a college, university, or technical school.
	___ B. To get a job.
	___ C. To enlist in the military.
	___ D. Other
Question 5:	Do you participate in one or more of the athletic programs at your school (basketball, football, soccer, hockey, tennis, volleyball, etc.)
	____ Yes (Y) ___ No (N)
Question 6:	Do you exercise daily?
	___ Yes (Y)___ No (N)
Question 7:	Do you spend at least 1 hour a week involved in an outdoor activity (walking, running, playing a game etc.)?
	___ Yes (Y)___ No (N)
Question 8:	Are you involved in any community service activity?
	____ Yes (Y) ___ No (N)

Figure 9.1: Survey questions developed by students at Rufus King High School

a survey they thought would address several important statistical questions related to the school's academic and extracurricular programs. A few of the survey questions are listed in Figure 9.1.

After students read the scenario, discuss the following questions:

1. Why do you think the students designing this survey wanted to know the grade level of a student completing the survey (Survey Question 2)?

Possible answer: *Several of the other questions might be answered differently based on a student's grade level. For example, is it possible 9th graders might be more or less involved in extracurricular activities than 12th graders? Does a student's plan after completing high school change over time? Might 12th graders answer Question 4 differently than 9th graders?*

2. Why do you think students included Survey Question 3? What was a possible reason to consider this question important?

Possible answer: *Survey Question 3 might be used to examine the growing interest in therapy pets to address stress or anxiety. If a program of this type were pursued, what type of pets would be selected? Do most students have a similar interest in the pets selected?*

3. Why might it be important to know if students exercise daily? Will most students understand what this question is asking? Will most students answer this question?

Possible answer: *Exercise may have different meanings to students. Some students may think of this as organized school activity. Other students may think of this as an individual activity involving walking, running, stretching, etc. The question was considered adequate by the Rufus King students, but whether they collected accurate responses was not clear. This question was unclear to some students answering the question and is a good example of how questions of this type have different interpretations. After the survey was distributed, students discussed that the following rewording of this question might have clarified some of the confusion: "Do you participate in at least 10 minutes a day of physical exercise either alone or as part of a group?"*

4. Why might it be important to know if students are involved in community service? Do students agree on what is meant by community service? Will most students answer this question?

Possible answer: *This question also needed more clarity. Community service was viewed by some students as service activities by the school; other students interpreted community service as an activity organized by other organizations or groups. Here again is an opportunity to discuss the importance of whether or not a question provided the intended information needed in the statistical study.*

Take the students through the process the Rufus King students followed.

Formulate a Statistical Question

The Rufus King students developed a series of statistical questions designed to provide a summary of the school's student population. Several possible statistical questions emerged from this project. For example, are students typically going to attend a college or university after high school or pursue other options? Are students likely to participate in the school's athletic programs? Do students typically spend at least one hour per week outdoors?

Discussion About Different Ways to Collect Appropriate Data

Explain that after the survey was designed and approved by the administration, a plan was needed to organize how students would complete the survey. It was possible, but not practical, to analyze completed surveys from more than 1,200 students enrolled in the high school. Students (under the direction of their teachers) discussed ways in which they might distribute surveys to obtain a sample that provided all students in the school the same opportunity to complete the survey.

Ask the students to read the four data collection options and answer the two questions for each option.

For each of the following four options, answer the two questions:

» Do you think this option will provide an accurate summary of the responses from students in the school?

» If this option is used, are there any groups of students who may not be represented? Explain your answer.

Option 1:

Consider placing computers at various locations around school (e.g., the cafeteria, library, computer lab) that are monitored by students from the mathematics class involved with this project. Students in the vicinity of the computers would be asked to complete the survey provided on the computer. After a student completed the survey, the students monitoring the computer would save the results and load a new survey for the next student to complete. At the end of the day, the responses from the completed surveys would represent the representative sample for analyzing the questions.

Option 2:

There are 35 students in the mathematics class involved with this project. Each member of the class would be encouraged to anonymously complete the survey. The completed surveys would comprise the representative sample for analyzing the questions.

Option 3:

Students in the mathematics class involved with this project would post the survey online using a service provided by a private company. Each member of the class would encourage friends to complete the survey, both through word of mouth and also through their social media accounts. The online service would provide completed surveys that comprise the representative sample for analyzing the questions.

Option 4:

Students enrolled in the mathematics class would distribute surveys both before or after school at various locations in the school building. At the end of the day, the completed surveys would comprise the representative sample for analyzing the questions.

After the students have read the four options and answered the two questions for each, have the students share their answers for each option.

Option: Place students in groups. Have each group create a poster that lists the pros and cons of each method. Also, list their choice for a method.

Discussion Points

Discuss with students that each of the options would provide a sample, but there would be questions as to whether a representative sample of the student population would have completed the surveys. In general, these options result in a *convenience sample*, or a sample that did not provide an opportunity for a cross section of the school's students to complete the survey. It is important to indicate that a statistical study based on a convenience sample, or any sample not representative of the population, may result in *bias* that raises questions regarding any conclusions of the study.

Note: Consider discussing with your students how they would collect a representative sample at their school. Would any of the previous options provide a representative sample? What other options might be considered?

Design and Implementation of a Plan to Collect Data

Hand out Student Worksheet 9.2 Data Collection Methods

Have the students read the Plan to Collect Data section on Student Worksheet 9.2.

Plan to Collect Data

The following is a summary of the plan implemented at Rufus King High School.

All students attending Rufus King are required to take an English course. Students involved in the survey project arranged providing the option of completing a survey during an English class with the school's English teachers. They estimated it would take fewer than five minutes to complete the survey. A specific day was identified to complete the survey. Students were also told by their English teachers that they did not have to complete the survey. Students involved in organizing this project provided an explanation of the project to the students several days before the survey was distributed by way of an all-school announcement. In addition, a flier was sent home to inform parents and guardians about the project.

The number of students who completed the survey was 1103.

Each survey was collected and given a specific identification number. Identification numbers from 1 to 1103 were assigned to the completed surveys. It was decided that 50 randomly selected surveys would form the sample for this study. Students generated 50 random numbers from 1 to 1103 using a graphing calculator. The 50 numbers generated by the calculator represented the 50 identification numbers and the 50 surveys selected to form the sample.

Ask your students to answer questions 1 to 5.

1. Do you think the above plan resulted in a sample that provided all students an equal chance to be selected in the sample? Explain you answer.

Possible answer: *Essentially, every student who was in attendance had an opportunity to complete the survey. This plan would result in a sample in which each completed survey had an equal chance of selection.*

2. Why do you think it was important to inform students about the project before they received the survey?

Possible answer: *It is important to emphasize that data of this type must convince students their time to complete the survey and their responses are important. If this study had been a research study conducted by professionals, a careful review of the survey questions would need to be conducted by a team of advisers. This team would also be responsible for evaluating details about the research project and how it would be communicated to students and their parents or guardians. Authentic statistical studies are held to high standards of communication and review.*

3. Why do you think it was important to inform parents and guardians about the project?

Possible answer: *Emphasize again the importance of maintaining communication in a statistical study. Also, students in high school are considered minors. Involving parents and guardians was an important requirement of the administration.*

4. Using the plan described, which students would not have completed the survey?

Possible answer: *Students absent from school or students who opted out of completing the survey would not have been included.*

5. Do you think the sample of 50 completed surveys represents a representative sample of all students?

Possible answers: *Encourage students to express their opinions to this question. It is anticipated students may comment that a sample of only 50 students would likely not be representative of the school's population. Students may also indicate a sample of only 50 surveys would not be large enough to obtain adequate summaries of the survey questions.*

Analyze the Data

Ask your students what type of data was collected for each of the eight survey questions.

Answer: *Categorical data.*

Ask students how they would summarize the responses to Question 1. What measure would they use to communicate what they have collected?

Possible discussion points: *Students may initially focus on the counts of Male responses or Female responses. Although the count of each category is important, it is the proportion of the* *number of males or the number of females to the sample size that will be the more important summary of the question in this statistical study. A similar proportion would be calculated for each of the other questions in the survey.*

Return to the discussion concerning whether the sample of 50 completed surveys represents a representative sample of all students. The main question is whether the selection of 50 surveys is large enough to estimate the proportions of the school population for each question. Will the proportion of females, proportion of students who exercise daily, or proportion of students who are involved in community service based on this sample of 50 students be the same as the school population?

The following example will help students understand that the random sample selected by students is likely to provide an adequate summary of the school population.

The sample of 50 students indicated 33 females and 17 males. A first step was to convert the number of females to a proportion—33/50 or 0.66 or 66% of the sample of 50 students was female and 17/50 or 0.34 or 34% of the students was male.

Share with your students the following summary of the school population posted on the school's website:

» Total enrollment (September 5): 1204 students

» Total number of females: 775

» Total number of males: 429

Based on this information summarizing the school population, the proportion of females at Rufus King High School at the time this project was conducted was 775/1204, or approximately 0.644 or 64%. The sample proportion of 66% was similar to the proportion of females of the school's population.

Interpret the Results in the Context of the Original Statistical Questions

Hand out Student Worksheet 9.3 Survey Results.

Direct students individually or in small groups to use the data presented on Student Worksheet 9.3 and answer Question 6.

Answers:

Q1 (Survey Question 1)

» Proportion of females: 33/50 or 0.66

» Proportion of males: 17/50 or 0.34

Q2 (Survey Question 2)

» Proportion of students in 9th grade: 15/50 or 0.30

» Proportion of students in 10th grade: 14/50 or 0.28

» Proportion of students in 11th grade: 16/50 or 0.32

» Proportion of students in 12th grade: 5/50 or 0.10

Q3 (Survey Question 3)

» Proportion of students who indicate they are a "dog person": 23/50 or 0.46

» Proportion of students who indicate they are a "cat person": 24/50 or 0.48

» Proportion of students who indicate they are neither: 3/50 or 0.06

Q4 (Survey Question 4)

» Proportion of students who plan to attend college after high school: 30/50 or 0.60

- » Proportion of students who plan to get a job after high school: 9/50 or 0.18
- » Proportion of students who plan to enlist in the military after high school: 6/50 or 0.12
- » Proportion of students who selected other: 5/50 or 0.10

Q5 (Survey Question 5)

- » Proportion of students who participate in the school's athletic program: 30/50 or 0.60

Q6 (Survey Question 6)

- » Proportion of students who exercise daily: 30/50 or 0.60

Q7 (Survey Question 7)

- » Proportion of students who spend at least one hour per week outdoors: 10/50 or 0.20

Q8 (Survey Question 8)

- » Proportion of students involved in community service: 23/50 or 0.46

Direct students to individually answer questions 7 to 12. After they have completed the questions, discuss these questions with the whole class.

7. Based on the above summaries, provide a brief description of the students attending this high school.

Summary answer: Students can focus on one or two summaries they find interesting about the school. For example, slightly more students considered themselves a cat person or only 20% of the students spend at least one hour outdoors. Remind students they are using a representative sample to describe the students in the school population.

8. What is your estimate of the *number of* students who participate in an athletic program from the total enrollment of 1204 students? Do you think your estimate is the exact number of students who participate in an athletic program?

Answer: *Assume the proportion of the school population participating in an athletic program is the same as the proportion of the sample. Therefore, an estimate of 0.60, or 60%, of the 1204 students is 722 students. This estimate is likely not the exact number of students who participate in an athletic program.*

9. Why might it be important to know the number of students and the proportion of students who participate in a school athletic program?

Possible answer: *Results could be used to evaluate the interest in an athletic program and whether the school's facility can effectively address the interest. Estimating the number of students involved in an athletic program might be used to determine whether the facilities (e.g., gyms or volleyball courts or bathrooms) are sufficient.*

10. What is your estimate of the students who participate in community service? Do you think your estimate is the exact number of students who participate in community service?

Answer: *Assume the proportion of the school population participating in community service is the same as the proportion of the sample. Therefore, an estimate of 0.46, or 46%, of the 1204 students is approximately 553.84 or 554 students. This estimate is likely not the exact number of students who participate in community service.*

11. Why might it be important to know the number and proportion of students who participate in community service?

Possible answer: *If the school is considering improving students' participation in community service, it is important to determine the current involvement.*

12. Why might it be important to know the proportion of students who spend at least one hour involved in outdoor activities?

Possible answer: *Several research studies have linked outdoor activity to student achievement. This survey question might be connected to other items (e.g., exercise, gender, grade level) that examine whether there are noticeable differences in outdoor activity based on these other categories.*

Exit Ticket

1. Based on the estimate of the relative frequencies from the surveys and the summary of the high-school enrollment of 775 females and 361 ninth graders, determine an estimate for each of the following two questions. For each question, indicate how you determined your estimate.

a. How many females do you think are involved in an athletic program at King?

Summary answer: *Assume the proportion females participating in an athletic program is the same as the proportion of students participating in an athletic program based on the sample. Therefore, 60% of the 775 female students, or 465 female students, is an estimate of the number of females who participate in an athletic program.*

b. How many 9th graders do you think are involved in at least 1 hour of outdoor activities per week?

Answer: *Assume the proportion of 9th graders involved in at least one hour of outdoor activities is the same as the proportion of students involved in at least one hour of outdoor activities from the sample. Therefore, assume 20% of the 361 9th grade students, or approximately 72.2 or 73 students, is an estimate. This also assumes the proportion of 9th grade students involved in outdoor activities is the same as the proportion of other grade levels.*

Exit Ticket Cont.

2. Consider the following data collection option:

   Students in the mathematics class involved with the project would number each table in the cafeteria. They would select 10 random tables at each lunch period and ask everyone sitting at the selected table to answer the survey.

   Do you think this option will provide an accurate summary of the responses from students in the school? If this option is used, are there any groups of students who may not be represented? Explain your answer.

Possible answer: *Since students usually sit with their friends at the same lunch table, it is likely students at the table would answer the survey questions in a similar manner. Students who do not eat in the cafeteria or go out or home for lunch are not represented.*

Further Explorations and Extension

Interest in completing a survey project at your school may also be considered a viable extension of this investigation. If the entire project could not be completed (due to time constraints or other challenges), designing a plan to carry out a project at your school that is similar to the one in this investigation might also be a valuable discussion and an exploration to consider.

Investigation 10

Is There an Association?

Summarizing Bivariate Categorical Data

Overview

The process a group of high-school students used to collect data from the student body was explored in Investigation 9. The students collected data using a survey, and many of the survey questions resulted in collecting data on categorical variables. In this investigation, students examine two of the survey questions and are asked to evaluate whether they think there is a connection or association between the two variables based on the conditional relative frequencies.

This investigation follows the four components of statistical problem solving put forth in the *Guidelines for Assessment and Instruction in Statistics Education (GAISE) Report.* The four components are formulate a question, design and implement a plan to collect data, analyze the data by measures and graphs, and interpret the results in the context of the original question. This is a GAISE Level B investigation.

The activities and several questions are also based on *Lessons from Probability Through Data*, published by the American Statistical Association (original copyright by Dale Seymour Publications, 1999) and available at *www.amstat.org/ASA/Education.*

Instructional Plan

Brief Overview

» Present Scenario 1 (no association).

» Develop a statistical question based on survey questions 5 and 7.

» Construct a two-way table summarizing survey results.

» Construct a row conditional relative frequency table based on survey results.

» Interpret the statistical question based on the row conditional relative frequency table.

» Present Scenario 2 (an association).

» Interpret the statistical question based on the row conditional relative frequency table.

Hand out Student Worksheet 10.1 Scenario 1 and Student Worksheet 10.3 Questions and Results.

Note: Students are presented with two scenarios. Scenario 1 examines two categorical variables that have no differences in the conditional relative frequencies. If there are no differences, then one variable does not suggest a possible connection to the second variable, or the two variables appear not to be connected. Scenario 2 explores a connection between two variables that results in noticeable differences in the conditional relative frequencies. The variables discussed in Scenario 2 indicate a possible association.

Scenario 1

A recent internet posting indicated a key factor in improving academic success for high-school students is to spend time outdoors. Students in the Rufus King project thought students involved in their school's athletic programs were more likely to spend time outdoors. To see if that was true, they incorporated a question into the survey (*www.health.harvard.edu/press_releases/spending-time-outdoors-is-good-for-you*).

After reading the scenario, discuss with students the following:

In your opinion, do students who participate in the organized sports program at their school and spend time outdoors have similar interests? Or, does participating in sports and spending time outdoors essentially have no connection?

Learning Goals

- » Calculate and interpret conditional relative frequencies from two-way frequency tables involving two categorical variables.
- » Evaluate whether the conditional relative frequencies are an indication of a possible association between the two categorical variables.

Mathematical Practices Through a Statistical Lens

MP2. Reason abstractly and quantitatively.

Statistically proficient students are able to summarize data to answer statistical questions. Students explain their summaries of data using relative frequencies and conditional relative frequencies.

Materials

Student worksheets are available at *www.statisticsteacher.org/statistics-teacher-publications/focus.*

- » Student Worksheet 10.1 Scenario 1
- » Student Worksheet 10.2 Scenario 2
- » Student Worksheet 10.3 Questions and Results (same as Student Worksheet 9.3)
- » Exit Ticket

Estimated Time

Two 50-minute class periods

Pre-Knowledge

- » Examine and evaluate a random sample of students' responses to a survey.
- » Summarize categorical data by relative frequencies and percent.
- » Estimate summaries of a population using relative frequencies of a sample.
- » Completion of Investigation 9.

Survey Response 1

Survey Number					Q5		Q7		
1					N		N		

Survey Response 2

Survey Number					Q5		Q7		
2					Y		Y		

Two of the questions on the Rufus King survey were the following:

» **Question 5:** Do you participate in one or more of the athletic programs at your school (basketball, football, soccer, hockey, tennis, volleyball, etc.)?

____ Yes (Y) ____ No (N)

» **Question 7:** Do you spend at least one hour a week involved in an outdoor activity (walking, running, playing a game, etc.)?

____ Yes (Y) ____ No (N)

Point out that there was confusion expressed about the wording of "outdoor activity" in survey Question 7. Discuss this question with your students. Ask them what they think the question is asking. The intent by the students involved with this project was to determine whether a student spent active time outdoors (running, walking, hiking, gardening). If students agree this question was not clear and possibly did not collect the intended information, discuss possible revisions. It is important to review survey questions both before they are distributed and after the information is collected to evaluate the goals of a statistical study.

Formulate a Statistical Question

Ask the students to consider the following statistical question as an investigation of participation in the sports program and spending time outdoors.

Statistical question: "Is there a connection between participation in an athletic program and spending time outdoors?"

Analyze the Data

Direct students to examine specific survey numbers on worksheet 10.3 Questions and Results.

Ask your students to describe the student who completed this survey. Discuss with them the question, "If most students who answered no to Question 5 and also answered no to Question 7, as in the Survey Response 1 table, do you think there is a connection between participating in an athletic program and spending time outdoors?" At this point in the discussion, students' responses to this question are opinions of what they think the connection might indicate.

Continue a similar discussion with the next examples:

» Discuss with students the question, "If most students who answered yes to Question 5 also answered yes to Question 7, as in the Survey Response 2 table, do you think there is a connection between participating in an athletic program and spending time outdoors?"

Survey Response 3

Survey Number				Q5		Q7		
3				Y		N		

Survey Response 10

Survey Number				Q5		Q7		
10				N		Y		

- » Discuss with students the question, "If most students who answered yes to Question 5 also answered no to Question 7, as in the Survey Response 3 table, do you think there is a connection between participating in an athletic program and spending time outdoors?"
- » Discuss with students the question, "If most students who answered no to Question 5 also answered yes to Question 7, as in the Survey Response 10 table, do you think there is a connection between participating in an athletic program and spending time outdoors?"

Ask your students if there are any other ways students could have answered these two questions. As students look through the rest of the sample (Student Worksheet 10.3), point out that each of the remaining surveys is represented by one of the previous four examples.

Explain that we would like to summarize all the survey results for these two questions. Refer the students to the empty table on Student Worksheet 10.1 Scenario 1 following the four survey responses. Discuss with the students the labels needed to complete the *column labels* representing Question 7 (or "Spend time outdoors" and "Do not spend time outdoors"). Continue the discussion by summarizing Question 5 with appropriate *row labels* (or "Participate in an athletic program" and "Do not participate in an athletic program"). Have the students add the labels to the table.

Answer: Table 10.1

Ask the students to answer questions 1 to 4.

1. Based on your table, in what cell would you count Survey Response 1?

Answer: Cell 5

2. Based on your table, in what cell would you count Survey Response 2?

Answer: Cell 1

3. Based on your table, in what cell would you count Survey Response 3?

Answer: Cell 2

4. Based on your table, in what cell would you count Survey Response 10?

Answer: Cell 4

In groups, have the students go through worksheet 10.3 Results and complete the frequency table provided by determining in what cells each of the other surveys would be counted.

Note: Consider advising students to use tally marks in cells 1, 2, 4, and 5 for each of the 50 surveys. After the 50 surveys have been tallied, complete Frequency Table 10.2.

Table 10.1

	Spend time outdoors	Do not spend time outdoors	Total
Participate in an athletic program	Cell 1	Cell 2	Cell 3
Do not participate in an athletic program	Cell 4	Cell 5	Cell 6
Total	Cell 7	Cell 8	Cell 9

Frequency Table 10.2

	Spend time outdoors	Do not spend time outdoors	Total
Participate in an athletic program	Cell 1 6	Cell 2 24	Cell 3 30
Do not participate in an athletic program	Cell 4 4	Cell 5 16	Cell 6 20
Total	Cell 7 10	Cell 8 40	Cell 9 50

Analyze the Data by Measures and Graphs

Introduce the vocabulary pertaining to the two-way table.

Point out to the students there are two types of cells represented in this table that should be discussed. The shaded cells (1, 2, 4, and 5) are called *joint cells*, as they record the number of students responding to specific categories of two questions. The other cells (3, 6, 7, 8, and 9) are called *marginal cells*, as they represent the margins of the table and record the number of students responding to a specific category of one question, with the exception of cell 9, which records the total number of surveys completed in this sample. If necessary, identify a cell and ask students to explain what that cell indicates. For example, cell 2 indicates the number of students who participate in an athletic program but do not spend time outdoors.

Ask your students to use the frequency table and answer questions 5 and 6.

5. What proportion of the survey-takers answered that they spend time outdoors *and* participate in an athletic program?

Answer: 6/50 = 0.12, or 12%

6. What proportion of the survey-takers answered that they do not spend time outdoors *and* do not participate in an athletic program?

Answer: 16/50 = 0.32, or 32%

Explain that the 0.12 and 0.32 are called *relative frequencies*. Relative frequencies are the proportions of the counts in each cell to the total number of surveys completed in the sample (or 50 surveys for this example). The relative frequencies provide a description of the proportion or percent of students in each cell to the total number of students in the sample.

Explain that while the relative frequencies summarize the results of the sample, they do not help determine whether the responses to the two questions are connected. The relative frequencies do not specifically examine the differences in the students who participate in an athletic program and do not participate in an athletic program.

Explain that to determine if there is a connection between participation in an athletic program and spending time outdoors, we need to find the proportion of those who participated in an athletic program who spent time outdoors and compare that to the proportion of those who did not participate in an athletic program who spent time outdoors.

Ask your students to highlight the first row of the frequency table by circling the entire row—those who participated in an athletic program.

Ask your students to answer questions 7 to 9.

7. Of the students in the survey, how many participated in an athletic program?

Answer: 30

8. Of those who participated in an athletic program, how many spent time outdoors?

Answer: 6

9. What proportion of students who participated in an athletic program spent time outdoors?

Answer: 6/30 = 0.20, or 20%

Explain that this value of 0.20 is called a *row conditional relative frequency*. It is the proportion of those who participated in an athletic program who spent time outdoors.

Direct your students to enter the 0.20 into cell 1 of Conditional Relative Frequency Table 10.3.

Ask your students to calculate the proportion of the 30 students who participated in an athletic program who did not spend time outdoors and enter this value into cell 2.

Answer: 24/30=0.80, or 80%

Ask your students to calculate the proportions in cells 4 and 5 of the students who spent time outdoors or did not spend time outdoors based on the condition they did not participate in the school's athletic program.

10. Ask your students to calculate the conditional relative frequencies for cells 3, 6, 7, 8, and 9.

The completed table is called a *row conditional relative frequency table* (See Table 10.3).

Optional: To help your students visualize the conditional relative frequencies of students who do and don't participate in athletics and spend time outdoors, demonstrate the construction of a segmented bar graph. Using the conditional relative frequency table, a segmented bar graph is shown in Figure 10.1. The conditional relative frequencies could also be visualized in side-by-side bar graphs.

Row Conditional Relative Frequency Table 10.3

	Spend time outdoors	Do not spend time outdoors	Total
Participate in an athletic program	Cell 1 6/30 = 0.20, or 20%	Cell 2 24/30 = 0.80, or 80%	Cell 3 30/30 = 1.00, or 100%
Do not participate in an athletic program	Cell 4 4/20 = 0.20, or 20%	Cell 5 16/20 = 0.80, or 80%	Cell 6 20/20 = 1.00, or 100%
Total	Cell 7 10/50 = 0.20, or 20%	Cell 8 40/50 = 0.80, or 80%	Cell 9 50/50 = 1.00, or 100%

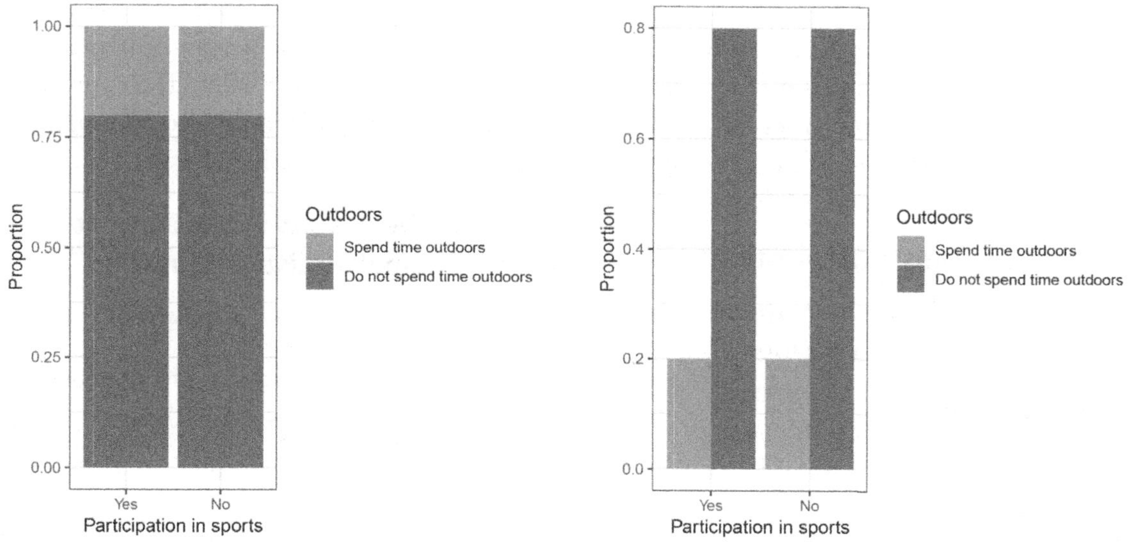

Figure 10.1: Segmented bar graph and side-by-side bar graph of athletic participation and time outdoors data

Interpret the Results in the Context of the Original Statistical Question

Ask your students to answer questions 11 to 17.

11. What is the proportion of students who spend time outdoors?

Answer: 0.20, or 20%

12. What is the conditional relative frequency of the 30 students in an athletic program who spend time outdoors?

Answer: 0.20, or 20%

13. What is the conditional relative frequency of the 20 students who do not participate in an athletic program who spend time outdoors?

Answer: 0.20, or 20%

14. If a student completing the survey participated in an athletic program, what is your estimate she or he spends time outdoors?

Answer: 0.20, or 20%

15. If a student completing the survey did not participate in an athletic program, what is your estimate she or he does not spend time outdoors?

Answer: 0.20, or 20%

16. Does knowing whether a student participates or does not participate in an athletic program change your estimate of spending time outdoors?

Answer: No

17. Do you think the questions about participating in an athletic program and spending time outdoors are connected? Explain your answer.

Possible answer: *The data do not indicate that participation in the athletic program and spending time outdoors are connected.*

Hand out Student Worksheet 10.2 Scenario 2

Scenario 2

Many schools, nursing homes, and residential communities are making therapy pets available to address anxiety and depression. Dogs and cats are most commonly used in pet therapy. However, fish, guinea pigs, horses, and other animals that meet screening criteria can also be used.

The type of animal chosen depends on the therapeutic goals of a person's treatment plan. In selecting a therapy pet for a person, there is debate as to whether females and males have different preferences in the selection of their therapy pets. *Source: www.healthline.com/health/pet-therapy*

Formulate a Statistical Question

Discuss with students an appropriate statistical question for this investigation.

Statistical Question

Is there a connection between gender and whether a person is a dog person or a cat person or neither a dog or cat person?

Analyze the Data

Frequency Table 10.4, involving the responses to item Survey Question 1 (Q1) and Survey Question 3 (Q3), is provided below.

Ask your students to answer Question 1.

1. Using the frequency table, complete the row conditional relative frequencies based on gender.

Answer: Row Conditional Relative Frequency Table 10.5

Optional: To help your students visualize the different percentages of females and males pet preferences, demonstrate the construction of a segmented bar graph. Using the conditional relative frequency table, a segmented bar graph is shown in figure 10.2. The conditional relative frequencies could also be visualized in side-by-side bar graphs.

Interpret the Results in the Context of the Original Statistical Question

Ask the students to answer questions 2 to 9.

2. What is the proportion of students who consider themselves a dog person?

Answer: 23/50 = 0.46, or 46%

3. What is the proportion of males who consider themselves a dog person?

Answer: 15/17= 0.882, or 88.2%

4. What is the proportion of females who consider themselves a dog person?

Answer: 8/33= *0.242, or 24.2%*

5. If a survey from the sample indicated the student completing the survey was male, what is your estimate he considers himself a dog person?

Answer: 0.882, or 88.2%

6. If a survey from the sample indicated the student completing the survey was female, what is your estimate she considers herself a dog person?

Frequency Table 10.4

	A. Dog Person	B. Cat Person	C. Neither	Total
Female	8	22	3	33
Male	15	2	0	17
Total	23	24	3	50

Row Conditional Relative Frequency Table 10.5

	A. Dog Person	B. Cat Person	C. Neither	Total
Female	8/33 = 0.242, or 24.2%	22/33 = 0.667, or 66.7%	3/33 = 0.091, or 9.1%	33/33 = 1.00, or 100%
Male	15/17 = 0.882, or 88.2%	2/17 = 0.118, or 11.8%	0	17/17 = 1.00, or 100%
Total	23/50 = 0.46, or 46%	24/50 = 0.48, or 48%	3/50 = 0.06, or 6%	50/50 or 1.00, or 100%

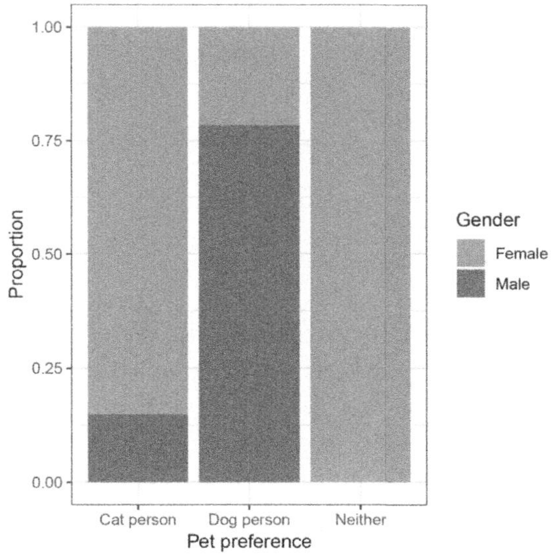

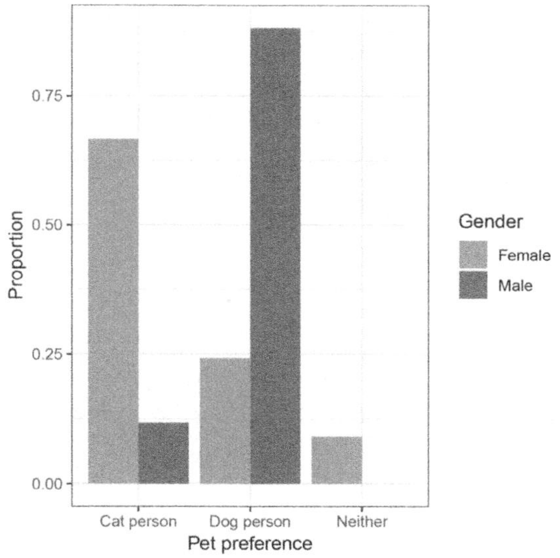

Figure 10.2: Segmented bar graph and side-by-side bar graph of females and males pet preferences data

***Answer:** 0.242, or 24.2%*

7. If a survey from the sample indicated the student completing the survey was male, what is your estimate he considers himself a cat person?

***Answer:** 2/17=0.118, or 11.8%*

8. If a survey from the sample indicated the student completing the survey was female, what is your estimate she considers herself a cat person?

***Answer:** 22/33=**0.667**, or 66.7%*

9. Do you think the responses to the question about animal preference are connected to gender? Explain.

***Possible answers:** The responses indicate there is a large difference in the proportion of males and the proportion of females who prefer a dog or a cat. There seems to be a connection in which males prefer dogs and females prefer cats.*

The differences in the conditional relative frequencies or proportions for each animal preference by gender indicates there is a possible connection. This connection is called an *association.*

Discuss with your students the general idea of association between two categorical variables.

Point out that the differences in conditional relative frequencies between two categorical variables indicate a possible association between the variables. The differences in the conditional relative frequencies indicate a connection is based on the conditions or responses to one of the survey questions.

It is important to remind students that whenever an association is noted in this type of statistical study, the connection suggested by the association is not *causal*. For this example, a person's gender does not cause the result of pet preference. Causation is important to analyze when examining experimental statistical studies.

Ask your students how might decisions that make therapy pets available for students be managed based on gender differences in pet preferences if there is a connection between gender and pet preference. As this is still a new challenge for schools, it is possible that students might suggest having both dogs and cats available is important (as opposed to just dogs or just cats).

Exit Ticket

Complete the following two-way frequency table for the variables of playing in the orchestra (Yes or No) and playing in the marching band (Yes or No) in which a sample of 50 students *indicates there is a possible association between the two variables.*

Why does your table indicate a possible association between the two questions involving participation in the orchestra and participation in the marching band?

	Play in the orchestra	Do not play in the orchestra	Total
Play in the marching band			20
Do not play in the marching band			30
Total	15	35	50

Possible Answer to the Exit Ticket

Answer will vary. As an example of how to evaluate answers, consider the following frequency table:

	Play in the orchestra	Do not play in the orchestra	Total
Play in the marching band	10	10	20
Do not play in the marching band	5	25	30
Total	15	35	50

Using the above frequency table, determine the conditional relative frequency table:

	Play in the orchestra	Do not play in the orchestra	Total
Play in the marching band	10/20 = 0.50, or 50%	10/20 = 0.50, or 50%	20/20 = 1.00, or 100%
Do not play in the marching band	5/30 = 0.167, or 16.7%	25/30 = 0.833, or 83.3%	30/30 = 1.00, or 100%
Total	15/50 = 0.30, or 30%	35/50 = 0.70, or 70%	50/50 = 1.00, or 100%

The conditional relative frequency table indicates there is a greater likelihood that if a survey is selected in which a student plays in the marching band, this student also plays in the orchestra than if a survey is selected in which the student does not play in the marching band and this student plays in the orchestra.

The rather large difference in the conditional relative frequencies indicates a possible connection or association of the variables involving orchestra and band. This particular example indicates there is a higher likelihood that if a student plays in the band, this student also plays in the orchestra.

Students should suggest values that add up to the marginal totals. Students should also select values that result in a noticeable or large difference in the conditional relative frequencies.

Further Explorations and Extensions

A blank template is provided for students to possibly explore two variables of their choice from the Rufus King High School data. (See Worksheet 10.3 Survey Questions Results.) Students should start by forming a statistical question based on selecting two questions from the survey. Students then organize the responses in a two-way frequency table. From the two-way frequency table, students create a relative frequency table and then a conditional relative frequency table. Based on the conditional relative frequency table, students would estimate whether the two variables they selected are possibly associated. They should summarize their statistical question based on the conditional relative frequencies.

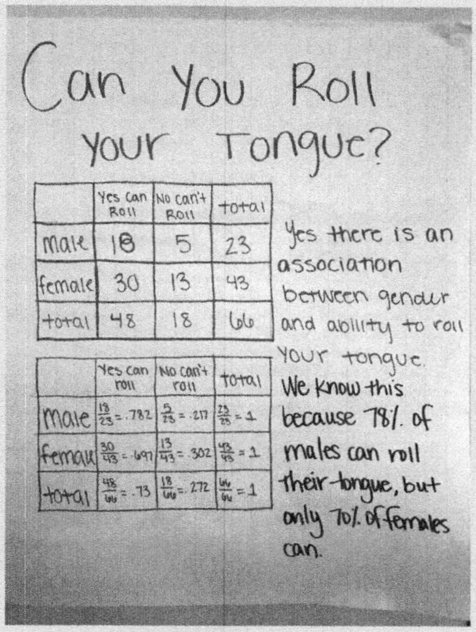

Sample of student work

Investigation 11

Independent or Not Independent Events?

Comparing Conditional Relative Frequencies

Overview

Investigation 10 explored whether there was an association between two categorical variables by examining the differences in conditional relative frequencies. In Part I of this investigation, students investigate the connection between conditional relative frequencies and independent events (in probability). In Part II, students design and conduct a simulation to determine if two events are independent.

This investigation follows the four components of statistical problem solving put forth in the *Guidelines for Assessment and Instruction in Statistics Education (GAISE) Report*. The four components are formulate a question, design and implement a plan to collect data, analyze the data by measures and graphs, and interpret the results in the context of the original question. This is a GAISE Level B activity.

The activity is based on lessons from *Probability Through Data* published by the American Statistical Association, available at *www.amstat.org/ASA/Education* (original copyright by Dale Seymour Publications, 1999).

Instructional Plan

Brief Overview

Part I

» Construct a row conditional relative frequency table based on year in school and win/lose a computer game.

» Develop the definition of independent events

Part II

» Formulate a statistical question on the likelihood a random sample of 100 will produce 27.5% success rate.

» Design and conduct a simulation.

» Answer the statistical question based on results of the simulation.

Part I: What Are Independent Events?

In Investigation 10, students determined whether two categorical variables on a set of randomly selected data were associated by calculating and comparing conditional relative frequencies of categorical variables. In the framework of probability, events are categories and conditional relative frequencies of categories are estimates for probabilities of events.

Part I is designed to provide an understanding of the definition of independent events.

Hand out Student Worksheet 11.1 Independent or Not Independent Events.

Ask students to read the scenario and answer questions 1 to 3. Then, discuss student answers to questions 1 to 3.

Scenario

Games can involve chance, skill, strategy, or some mixture of them. This investigation is interested in games of chance such as Candy Land, Chutes and Ladders, and the card game War.

1. Identify a game that determines a win or loss by chance alone. Explain how chance

is involved and how skill or strategy are not involved.

2. Identify a game in which a win or loss is primarily determined by the skill of a player or players. Explain.
3. Identify a game in which a win or loss involves both chance and skill or strategies.

Have your students read Game: Over or Under.

Game: Over or Under

The computer science students at Rufus King High School designed a game to be played on a computer they call Over or Under. The directions were provided in the opening screen.

Learning Goal

Understand and interpret the connection between conditional relative frequencies and independent events in probability and the definition of independent events.

Mathematical Practices Through a Statistical Lens

MP2. Reason abstractly and quantitatively.

Statistically proficient students are able to summarize data to answer statistical questions. Students explain their summaries of data using relative frequencies and conditional relative frequencies as probabilities. Students interpret relative and conditional relative frequencies to reason about the population from which a sample was selected.

Materials

Student worksheets are available at *www.statisticsteacher.org/statistics-teacher-publications/focus.*

- » Student Worksheet 11.1 Independent or Not Independent Events
- » Student Worksheet 11.2 Simulation
- » Student Worksheet 11.3 Template for Conducting the Simulations (on stock paper)
- » For the simulation component of this investigation (Part II), students will need a small paper bag (one per small group) and the cut-out slips of paper from the template provided with this investigation.
- » Exit Ticket

Estimated Time

Two 50-minute class periods. Part I develops an understanding of independent events. This section is expected to take one class period. Part II directs students in conducting a simulation of the scenario in the investigation.

Pre-Knowledge

Summarize data by relative frequencies, conditional relative frequencies, and percentages (completion of Investigation 10).

```
Each one of the numbers 0, 1, 2,
3, 4, and 5 is behind the fol-
lowing cards labeled A, B, C,
D, E, and F. They are in random
order. Each number is used with
no repeats. Click on any three
cards. If the sum of the numbers
behind the cards is 6 or less,
then you win the game. If the
sum of the numbers is greater
than 6, then you lose. Hit the
Start icon to begin the game.
Have fun!

Start
```

Justin played the game. The following opening screen starts the game.

Card A	Card B	Card C	Card D	Card E	Card F
?	?	?	?	?	?

Justin clicked on cards A, C, and D. The next screen indicated the following:

Card A		Card C		Card D		
3	+	0	+	2	=	5

You WIN!

If the sum had been greater than 6, Justin would have lost the game.

The computer science students decided to test out their game to determine if it would interest the students in their school. They were given permission to randomly select 100 students from their school and ask them several questions, including if they would play their game, what year in high school they were in (1st, 2nd, 3rd, or 4th), and whether the game was interesting. Each of 100 selected students agreed to play the game once and record whether they won or lost.

Table 11.1

	Number of students who won the game	Number of students who lost the game	Total number of students who played the game
1st- or 2nd-year Student	11	29	40
3rd- or 4th-year Student	19	41	60
Total	**30**	**70**	**100**

Exactly 100 students played the game once. Table 11.1 summarizes the results.

Students in the computer science class wanted to investigate if winning the game was connected to grade level. Are 3rd- or 4th-year students better at playing games of chance than 1st- or 2nd-year students?

Have your students complete Question 4.

Note: This question requires students to have completed Investigation 10.

4. Complete the conditional relative frequency table 11.2 of winning or losing the game categories based on year of student.

Answer: *Table 11.2*

Explain to your students that we are now going to use conditional relative frequencies of categories as estimates for probabilities of events. For example, what is the probability a randomly selected student wins the game?

Answer: *30/100=0.30, or 30%*

Have your students complete Question 5.

5. Use the conditional relative frequencies as estimates of conditional probabilities and complete Table 11.3, the conditional probability of events.

Table 11.2

	Conditional Relative Frequencies for Wins Based on Year	Conditional Relative Frequencies for Losses Based on Year	Totals
1st- or 2nd-Year Student	11/40 = 0.275, or 27.5%	29/40 = 0.725, or 72.5%	40/40 = 1.00, or 100%
3rd- or 4th-Year Student	19/60 = 0.317, or 31.7%	41/60 = 0.683, or 68.3%	60/60 = 1.00, or 100%
Totals	30/100 = 0.30, or 30%	70/100 = 0.70, or 70%	100/100 = 1.00

Table 11.3

	Conditional Probability of Winning Based on Year	Conditional Probability of Losing Based on Year	Totals
1st- or 2nd-Year Student	11/40 = 0.275, or 27.5%	29/40 = 0.725, or 72.5%	40/40 = 1.00, or 100%
3rd- or 4th-Year Student	19/60 = 0.317, or 31.7%	41/60 = 0.683, or 68.3%	60/60 = 1.00, or 100%
Totals	30/100 = 0.30, or 30%	70/100 = 0.70, or 70%	100/100 = 1.00

Answer: Table 11.3

Have your students answer questions 6 to 8.

Interpret the table of conditional probabilities.

6. If winning this game is totally based on chance and not connected to the year a student is in school, what is the probability that a randomly selected student wins the game?

Answer: 30%

7. If winning this game is totally based on chance, what is the conditional probability that a 1st- or 2nd-year student would win the game?

Answer: 27.5%

8. If winning the game is totally based on chance, what is the conditional probability that a 3rd- or 4th-year student would win the game?

Answer: 31.7%

Note: You may want to explain to students that this probability is an empirical estimate. A theoretical probability could also be derived. However, for this investigation, the probability based on the above sample will be considered the probability of winning the game.

Discuss with students the following definition of *independent events.*

> Two events are *independent* when knowing that one event has occurred does not change the likelihood that the second event will occur.

Have the students answer Question 9.

9. If event A is "winning the game" and event B is "1st- or 2nd-year student," are A and B independent events?

Answer: *The probability that a randomly selected student wins the game is 30%. The probability that a 1st- or 2nd-year student wins the game is 27.5%, so winning the game is not independent of being a 1st- or 2nd-year student.*

Note: The conditional probability of winning the game for a 3rd- or 4th-year student is 31.7%. The conclusion is there is a higher probability that a student in their 3rd or 4th

Table 11.4

	Number of Students Who Won the Game	Number of Students Who Lost the Game	Total Number of Students Who Played the Game
1st- or 2nd-Year Student	12	28	40
3rd- or 4th-Year Student	18	42	60
Totals	30	70	100

Table 11.5

	Conditional Probability of Winning	Conditional Probability of Losing	Totals
1st- or 2nd-Year Student	12/40 = 0.30	28/40 = 0.70	40/40 = 1.00, or 100%
3rd- or 4th-Year Student	18/60 = 0.30	42/60 = 0.70	60/60 = 1.00, or 100%
Totals	30/100 = 0.30, or 30%	70/100 = 0.70, or 70%	100/100 = 1.00

year will win the game as compared to one in their 1st or 2nd year.

Direct your students to complete questions 10 and 11 based on the hypothetical results in Table 11.4.

10. Using the hypothetical results in Table 11.4, complete the row conditional probability of events based on year in school in Table 11.5.

Answer: Table 11.5

11. Using the hypothetical results, if event A is "winning the game" and event B is "1st- or 2nd-year student," are events A and B independent events?

Answer: The probability that a randomly selected student wins the game is 30%. The probability that a 1st- or 2nd-year student wins the game is 30%, so winning the game is independent of being a 1st- or 2nd-year student.

Part II: Using Simulation to Determine if Two Events Are Independent

Note: Part II simulates many samples of size 100 from a population of all Rufus King High School students in which the two events are assumed to be independent and uses that distribution to determine how likely a random sample of 100 students would produce 27.5% of 1st- or 2nd-year students winning the game. Recall that 27.5% was the observed percentage found in Investigation 10.

Refer your students to the sample the computer science students took. The row conditional relative frequency table based on year in school is shown in Table 11.6.

Ask your students to help complete a row conditional relative frequency (Example shown in Table 11.7) based on the year in school.

Table 11.6

	Number of students who won the game	Number of students who lost the game	Total number of students who played
1st- or 2nd-year Student	10	30	40
3rd- or 4th-year Student	20	40	60
Total	30	70	100

Draw a number line on the board and record the proportion of 1st- or 2nd-year students who would win the game.

Example:

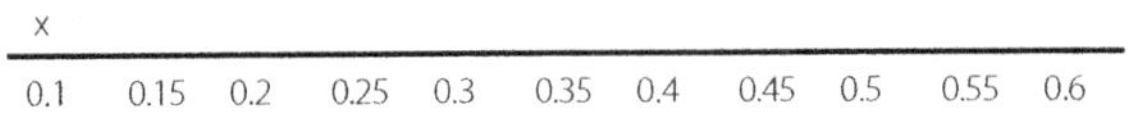

Ask your students:

What does the 0.25 represent?

Answer: *Assuming the two events are independent, 25% of the 1st- or 2nd-year students won the game.*

Based on the sample result of 25%, are you convinced the two events are independent?

Explain that we need a large number of simulation results to draw a conclusion.

Place students in small groups. Each group will require a bag of the slips of paper cut out from the template. Have your students refer to Worksheet 11.2 Simulation.

Simulation Steps per Trial:

Step 1: Thoroughly mix the slips of paper in the paper bag.

Step 2: Pick 30 slips representing the students who won the game.

Step 3: Count the number of slips that have a 1 on them.

Step 4: Determine the estimated probability of a 1st- or 2nd-year student winning the game and record the estimated probability on a data recording sheet similar to the following:

Trial number	Number of slips representing 1st- or 2nd-year students winning the game	Probability estimate that a 1st- or 2nd-year student wins the game
Example	10	10/40 = 0.25
1		
2		
3		
4		
5		

Step 5: Repeat steps 1 to 4 at least four more times (for a total of five trials). Record each trial result on the recording sheet.

After each small group of students has collected these data for at least five trials, direct each group to add their results to the dot plot on the board.

Note: If more trials are needed, use the 25 simulations in Table 11.8 obtained by the above process.

Figure 11.1 is the dot plot of the 25 simulations.

Option: Direct students to obtain the results of the simulation from an applet or

Table 11.7

	Conditional Probability of Winning	Conditional Probability of Losing	Totals
1st- or 2nd-Year Student	10/40 = 0.25	30/40 = 0.75	40/40 = 1.00
3rd- or 4th-Year Student	20/60 = 0.33	40/60 = 0.66	60/60 = 1.00
Totals	30/100 = 0.30	70/100 = 0.70	100/100 = 1.00

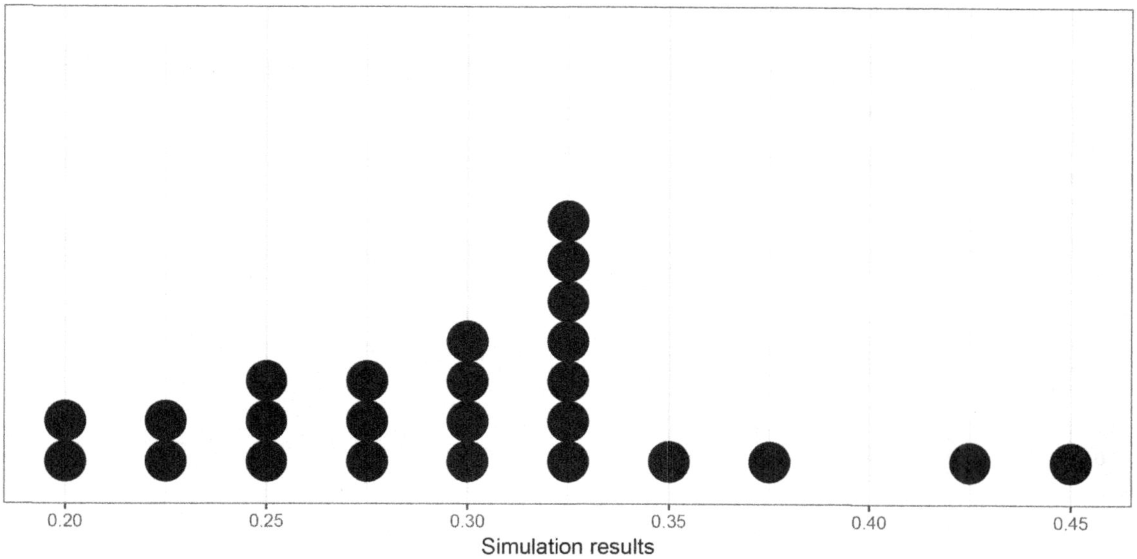

Figure 11.1: Dot plot of the 25 simulations

statistical software application like StatKey (*www.lock5stat.com/StatKey*).

Interpret the Results in the Context of the Original Question

Ask your students to answer questions 12 and 13 that estimate the likelihood of 1st- or 2nd-year students winning the game if the events are independent.

12. Based on the class dot plot of the simulated probabilities, what estimates of the proportion of a 1st- or 2nd-year student winning the game are most likely to occur under the assumption that the probability that 1st- or 2nd-year students win the game is 30%? Explain your answer.

Answer: *Students would identify the probabilities that occurred the most from the simulations. For the dot plot of 25 simulations, answers such as 0.25 to 0.32 would be expected. Note that most of the simulations were within that interval. As the number of simulations are added to the dot plot, the build-up of the values around 0.3 is more pronounced.*

13. Do you think the sample of 100 students collected by the computer science students could have come from a population in which the events of grade level and winning the game are independent? Explain your answer.

Answer: *A proportion 27.5% representing the probability of 1st- or 2nd-year students winning the game fits within the interval describing most of the expected proportions from the simulations. As a result, this sample is likely to have been drawn from a population in which the events are independent, or nearly independent.*

Table 11.8 Twenty-Five Simulations

Trial	1	2	3	4	5
Probability of a 1st- or 2nd-Year Student Winning	10/40 = 0.25	12/40 = 0.300	13/40 = 0.325	11/40 = 0.275	13/40 = 0.325
Trial	6	7	8	9	10
Probability of a 1st- or 2nd-Year Student Winning	13/40 = 0.325	15/40 = 0.375	13/40 = 0.325	8/40 = 0.200	13/40 = 0.325
Trial	11	12	13	14	15
Probability of a 1st- or 2nd-Year Student Winning	14/40 = 0.350	11/40 = 0.275	18/40 = 0.450	12/40 = 0.300	12/40 = 0.300
Trial	16	17	18	19	20
Probability of a 1st- or 2nd-Year Student Winning	17/40 = 0.425	9/40 = 0.225	13/40 = 0.325	12/40 = 0.300	10/40 = 0.250
Trial	21	22	23	24	25
Probability of a 1st- or 2nd-Year Student Winning	9/40 = 0.225	8/40 = 0.200	10/40 = 0.250	11/40 = 0.275	13/40 = 0.325

Exit Ticket

The students selected in the original sample collected by the computer science class also answered the question of whether a student participates in the school's extracurricular activities. Based on this sample, do you think the events of participation in a school's extracurricular activities and grade level are likely to be independent events in the population? Explain your answer.

	Participates in the School's Extracurricular Activities	Does Not Participate in the School's Extracurricular Activities	Total
1st- or 2nd-Year Student	34	6	40
3rd- or 4th-Year Student	6	54	60
Totals	40	60	100

Answer: *Students determine the conditional probabilities using the year in school of the students. (See table below.) The probability of 85% that a 1st- or 2nd-year participates in extracurricular activities is quite different than the 40% for all students participating in the school's extracurricular activities. This major difference indicates that the assumption the events are independent is not likely to be accurate. If a student is selected who is a 1st- or 2nd-year student, the estimate this student participates in extracurricular activities is higher than if the student selected was a 3rd- or 4th-year student.*

	Participates in the School's Extracurricular Activities	Does Not Participate in the School's Extracurricular Activities	Total
1st- or 2nd-Year Student	34/40 = 0.85, or 85%	6/40 = 0.15, or 15%	40/40 =1.00, or 100%
3rd- or 4th-Year Student	6/60 = 0.10, or 10%	54/60 = 0.90, or 90%	60/60 = 1.00, or 100%
Totals	40/100 = 0.40, or 40%	60/100 = 0.60, or 60%	100/100 = 1.00, or 100%

Extension

Close? Close Enough?

The conclusion in this investigation was the sample was likely to have been drawn from a population in which the events are independent, or nearly independent. We observed 27.5% for 1st- or 2nd-year students who won the game, and this is relatively close to 30%.

The probabilities in this investigation were estimated empirically based on a random sample of 100 students. Students were concerned, however, that the probabilities were not equal, and although within an interval that included most of the probabilities from a simulation, were the probabilities close enough to conclude they were independent events? What is "close enough"? It was noted by students that rarely will probabilities derived from the sample be equal.

Investigation 18, "How Stressed Are You?" investigates a similar statistical question in which the difference between two proportions based on two categories are close enough to conclude there is no significant difference in the categories. When is the difference close enough to indicate the categories are not significantly different? For now, students will use their best judgment to estimate whether they think the probabilities are close enough based on a comparison to simulated probabilities. Acknowledge, however, that estimating whether the probabilities are close is important and will be more precisely defined. Topics involving p-values or confidence levels will be addressed as they continue their study of statistics. The answers to these questions are especially important as students move to a more precise study of inferential statistics.

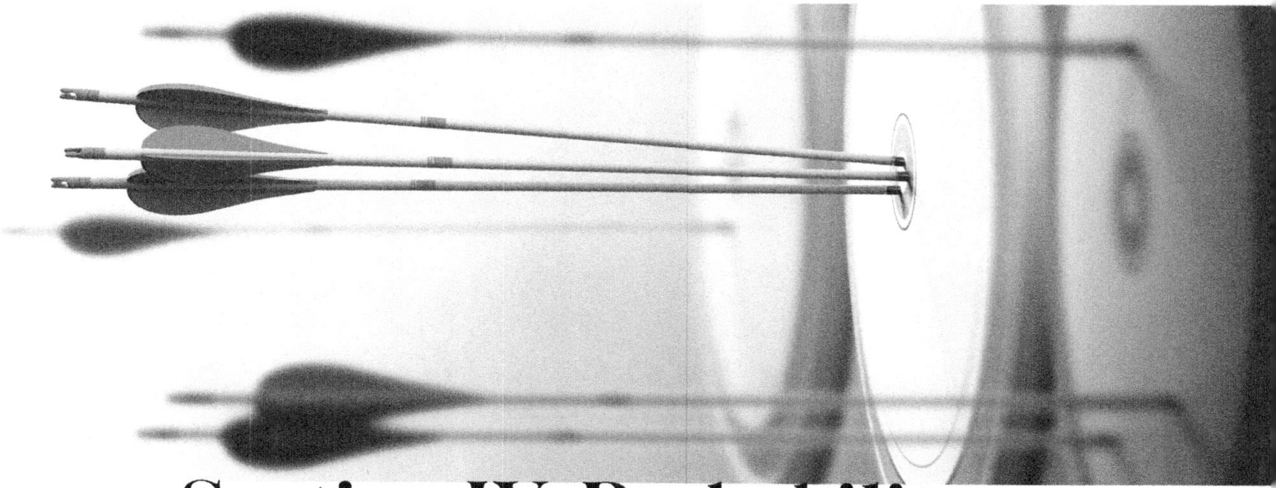

Section IV: Probability

Investigation 12

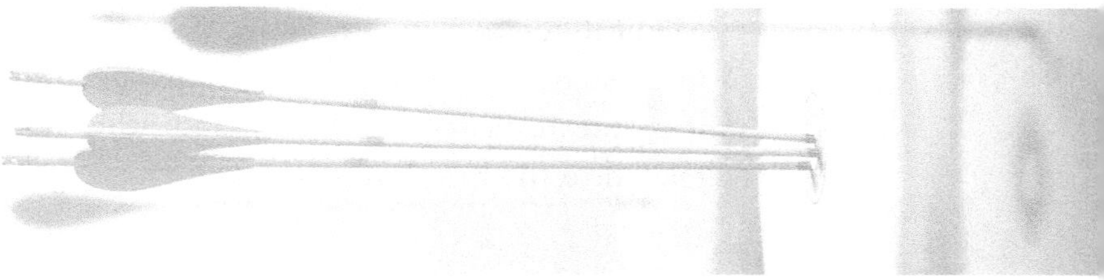

Chances of Getting the Flu?
Simulations

Overview

This investigation develops a probability distribution through the design and use of a simulation. It follows the four components of statistical problem solving put forth in the *Guidelines for Assessment and Instruction in Statistics Education (GAISE) Report*. The four components are formulate a statistical question, design and implement a plan to collect data, analyze the data by measures and graphs, and interpret the results in the context of the original question. This is a GAISE Level B activity.

This activity is based on a simulation problem from *The Art and Techniques of Simulation*, published by Dale Seymour and the American Statistical Association. (This module is part of the Quantitative Literacy Series. Though out of print, the book is available through book resale sites and on Amazon.com.)

Instructional Plan

Brief Overview

» Read and discuss the scenario about the spread of flu in an apartment building.

» Formulate the statistical/probabilistic question: "What is an estimate for the probability that all six people who live in an apartment building will get the flu?"

» Demonstrate the steps to conduct a simulation to answer the probabilistic question.

» Have students conduct the simulation using a die or technology and report their results.

» Collect class data in a table, convert the results to relative frequencies and a probability distribution.

» Use the probability distribution to answer the statistical/probabilistic question.

Hand out Student Worksheet 12.1 Flu Epidemic. Direct your students to read the first paragraph in the scenario.

Scenario

Did you get a flu vaccine last year? If so, did you still get the flu?

Infectious diseases (or diseases that are often caused by a bacteria or virus) are extensively researched in the medical field. These diseases result in colds, seasonal flu, and major epidemics that affect large numbers of people or animals in some cases.

In the fall of 1918, a flu pandemic erupted and became one of the greatest loss of lives the world had ever seen. By many accounts, the flu claimed between 2.5% and 5% of the global population. At that time, there was no flu vaccine, no antiviral drugs, and no antibiotics to help lessen the number of patients who got the flu or aid in the recovery from the flu.

As a result of this pandemic, countries began to put a greater emphasis on the study of patterns, causes, and effects of diseases.

Medical researchers are actively involved in understanding what causes the disease, how it is spread, how long it lasts, and other data related to the health of patients.

Source: www.smithsonianmag.com/history/how-1918-flu-pandemic-revolutionized-public-health-180965025

Learning Goals

- Design and carry out a simulation to estimate the probability of a random event.
- Develop a non-uniform probability distribution based on a simulation.

Mathematical Practices Through a Statistical Lens

MP5. Use appropriate tools strategically.

Statistically proficient students are able to use technological tools to carry out simulations for exploring and deepening their understanding of statistical and probabilistic concepts.

Materials

Student worksheets are available at *www.statisticsteacher.org/statistics-teacher-publications/focus.*

- Large foam die
- Die for each pair of students
- Technology like the TI-84 graphing calculator with ProbSim app or a similar rolling die application such as *www.random.org/dice*
- Student Worksheet 12.1 Flu Epidemic
- Student Worksheet 12.2 Simulation Steps
- Exit Ticket

Estimated Time

One 50-minute class period

Pre-Knowledge

Students should already be able to find the probability of simple events.

Students understand the probability of an event E is equal to:

$$P(E) = \frac{\text{Number of trials favorable to E}}{\text{Total number of trials in the experiment}}$$

Discuss with students the flu scenario and ask what type of precautions they can take to avoid getting the flu.

Ask your students to read the flu example.

Flu Example

Consider the following simple example of an infectious disease, like a cold or flu, and how it spreads throughout a small apartment building.

Suppose a strain of the flu has a one-day infection period (i.e., a person with the flu can only infect another person for one day and, after that day, the person can't spread the flu and is immune—that is, once you get the flu, you can't get this strain of flu again). This strain of flu is potent; if a person comes into contact with someone with the flu, that person will get the flu for certain.

Six people live in a small apartment building. One person catches this very infectious strain of flu and randomly encounters one of the other tenants during the infection period, and this second tenant gets this strain of flu. This second tenant infected with the flu visits a third tenant at random during the next day, and this third tenant gets the flu. The process continues with a newly infected person randomly visiting someone who hasn't had the flu or visiting an immune person and the strain of flu dies out. If an infected person visits an immune person, then the spread of the flu will end, as the flu in this example has only a one-day infection period.

Ask your students to summarize how this strain of flu spreads.

What is the least number of tenants who could get the flu?

Answer: *Two tenants, The first tenant gets the flu and visits a second tenant, who then goes back and visits the first tenant.*

What is the highest number of tenants who could get the flu?

Answer: *All six tenants*

Formulate a Statistical Question

Discuss with your students that one way to investigate an estimate of the number of people who would get the flu in this apartment building is to design and conduct a simulation. A simulation is a procedure developed for answering questions about real problems by running experiments that resemble the real-life situation. Instead of finding a large number of apartment buildings with six apartments and one person with the flu, a simulation could be designed to provide outcomes of the number of people who get the flu.

Ask students to consider the statistical/probabilistic question: "What is an estimate for the probability that all six people who live in an apartment building will get the flu?"

Collect Appropriate Data

To help your students understand the scenario, conduct a simulation involving them.

» Select six students and have them come to the front of the room. These six students represent the people living in the apartment building. Number each student from 1 to 6.

» *Day 1:* Roll the large foam die to determine Patient Zero, who will have the flu first. For example, if a 3 is rolled, then Person 3 has the flu. Have Person 3 roll the die, and then have Person 3 visit the person whose number is rolled. For example, a 4 is rolled. Remember this flu is potent; if a person is "visited," they will get the flu. Now two people have gotten the flu—persons 3 and 4. If Person 3 rolled a 3, then Person 3 would roll again since a person can't visit him/herself.

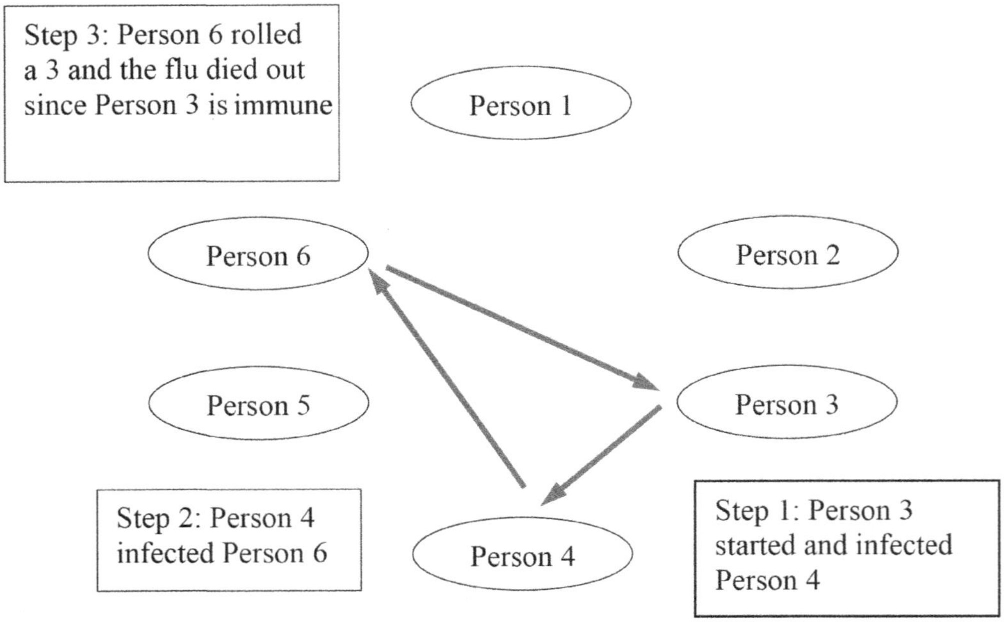

Figure 12.1

- *Day 2:* Person 3 is now immune (once you have had the flu, you can't get it again and you are no longer contagious) and Person 4, who now has the flu rolls a die and visits (infects) the person whose number was selected. For example, Person 6. Three people (3, 4, and 6) now have had the flu, unless Person 4 was to roll a 3. In that case, the flu would die out since the infected person visited a person who already had the flu. If the person rolls his/her own number, have the person roll again since a person can't visit him/herself.

- *Day 3:* Person 3 and Person 4 are immune. Person 6, who now has the flu, rolls a die and visits a person. Continue until a person visits someone who has already had the flu (i.e., immune) or someone who has not been infected. If the person rolls his or her own number, have the person roll again since a person can't visit him or herself.

Make a note of the number of people who got the flu.

Note: Students could also draw six circles—one for each person in the apartment building—and draw lines connecting the circles to show how the flu spreads as the simulation progresses.

Figure 12.1 illustrates the example above showing one trial in which three people were infected before the flu died out.

Emphasize that the goal is to design and conduct a simulation to find an estimate for the probability that all six people living in an apartment building will get the flu.

Share (consider posting) the steps to designing and conducting a simulation. Student Worksheet 12.2 Simulation Steps lists the steps.

Steps

1. State the problem or statistical/probabilistic question.
2. Define the simple events that form the basis of the simulation.
3. State any underlying conditions that need to be made so the answer to the probabilistic question can be determined.
4. Decide on a model that will be used to match the probabilities. Describe how random numbers will be assigned to match the probabilities described in the problem. Determine what constitutes a trial and what will be recorded.
5. Conduct the first trial.
6. Record the results of the trial.
7. Continue to run trials. Run a large number of trials. Remember to report the result of each trial.
8. Summarize the results of the trials and draw conclusions.

Go through the steps for this simulation using a die or large foam die.

1. State the problem (probabilistic question) so the objective of the simulation is clear. *What is an estimate for the probability all six people living in an apartment building will get the flu?*
2. Define the simple events that form the basis of the simulation. *Infected person randomly visits another person in the apartment building. If a person is randomly visited, they will get the flu, unless they have already had the flu.*
3. State any underlying conditions that need to be made so the answer to the probabilistic question can be determined.

Table 12.1

Trial Number	Who Was Infected (# on Each Roll	Number of People Infected
1	3,4,2,5,3	4
2	6,6,2,6	2
3		

Conditions: Visits are done randomly. Only one person can become infected at a time. Person can infect others for only one day.

4. Decide on a model that will be used to match the probabilities. Describe how random numbers will be assigned to match the probabilities described in the problem. Determine what constitutes a trial and what will be recorded. *Number the people from 1 to 6. Roll a die to simulate the visit by the infected person. (Persons can't visit themselves.) A trial is rolling the die until the flu dies out—a person with the flu visits someone who is immune (already had the flu). The number of people infected will be recorded.*
5. Define and conduct the first trial. *The first roll of the die determines which person was the first person to get the flu. Continue to roll the die until whoever is the current infected person visits an immune person (someone who has already had the flu). That is, roll until a number (other than the infected person's) is repeated. The trial is then over.*
6. Record the results of the trial. *Record the trial number, the results of each roll, and the number of people infected in a table, as shown in Table 12.1.*
7. Continue to run several more trials. Remember to record the result of each trial. *Repeat steps 5 and 6 a large number of times (at least 50 for the class). Give each pair of*

Table 12.2

Number of People Infected	Frequency
2	
3	
4	
5	
6	
Total	

Table 12.3

Number of People Infected	Frequency
2	17
3	33
4	22
5	8
6	2
Total	82

students a die and have them conduct at least five trials and collect the class results in a table.

Explain that an accurate estimate for a probability requires that a large number of trials be conducted (at least 50 for the whole class). Divide the students into groups of two. One person rolls the die and the other records the outcomes in a chart. Ask each group of students to conduct at least five trials.

After the groups have completed at least five trials, collect each group's results in Table 12.2. You are collecting the number of people infected for each trial.

Sample results from a class of 9th graders are shown in Table 12.3.

Option: Demonstrate how to use technology (e.g., ProbSim app on TI-84 Graphing calculator or other rolling die simulator) to collect a large number of trial results.

Analyze the Data

After the simulation has been run for a large number of trials and the results collected in a table, ask the students to answer questions 1 to 4.

1. Fill in Table 12.2 using the class simulation results.
2. Construct a dot plot of the class simulation results.

Possible answer: *Sample results from a 9th grade class in Figure 12.2*

3. What is the most likely number of people living in the apartment building who will get the flu?

Possible answer: *Three people*

4. Add a column to Table 12.3. Label the column Relative Frequency. Complete the relative frequency column in Table 12.4.

Answer: *Based on the example*

Explain that Table 12.4 gives estimates for the relative frequency of various successes (the number of persons who become infected). The relative frequencies for the different number of successes can be thought of as the probability of the number of successes. This table describes a *probability distribution.*

Let X = Number of people infected and P(X) = the probability of x people being infected.

5. What is an estimate for the probability that all six people living in an apartment building will get the flu?

Answer (based on the example in Table 12.5): *0.024, or 2.4%*

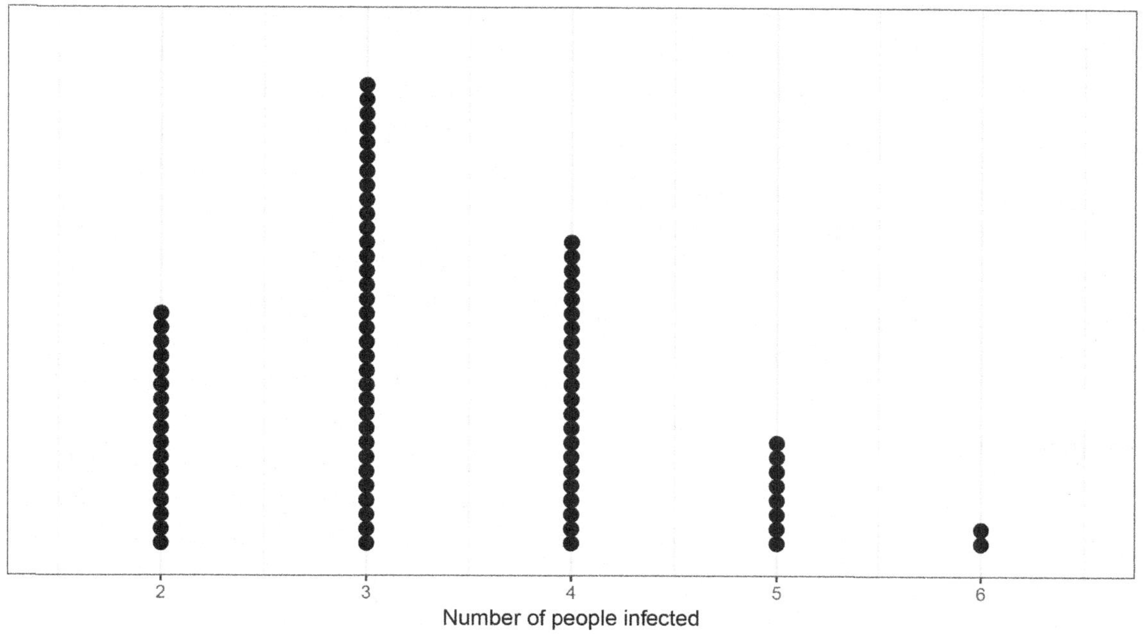

Figure 12.2: Dot plot of the class simulation results

Interpret the Results in the Context of the Original Question

Ask students to answer this question based on the simulation model they designed and conducted.

6. How did you model the spread of the flu in the apartment building? And how did you use this model to find an estimate for the probability that all six people living in the apartment building will get the flu?

Possible answer: *We modeled the spread of the flu by using a six-sided die. Each side of the die represented one person in the apartment building. We rolled the die and recorded the person who got the flu. We continued until a person visited someone with the flu, which caused the flu to die out. We recorded the number of people who got the flu and repeated the simulation a large number of times. After many trials, we were able to estimate the probability of all six people getting the flu as 2.4%*

Summary

To help summarize this simulation, ask your students the following questions:

7. What model could be used if there were eight people in the apartment building?

Table 12.4

Number of People Infected	Frequency	Relative Frequency
2	17	17/82 = 0.207
3	33	33/82 = 0.402
4	22	22/82 = 0.268
5	8	8/82 = 0.098
6	2	2/82 = 0.024
Total	82	1.0

Table 12.5

X	P(X)
2	17/82 = 0.207
3	33/82 = 0.402
4	22/82 = 0.268
5	8/82 = 0.098
6	2/82 = 0.024
Total	1.0

Possible answers: *An eight-sided die, randomly selecting numbers from 1 to 8 from a hat or bag, random number generator on computer or calculator*

8. How do you think the probability of all eight people in an apartment building getting the flu compares with the probability of all six people getting the flu?

Answer: The probability of eight would be smaller than the probability of six getting the flu.

Additional Ideas

1. Design and conduct a simulation for the following problem: The chance of contracting strep throat when encountering an infected person is estimated as 0.15. Suppose the four children of a family encounter an infected person. Conduct a simulation to estimate the probability of at least two of the children getting strep throat. State the conditions needed to simulate the problem.
2. A high-school algebra teacher has eight keys, but she never recalls which one fits her office door lock. She tries one key at a time, each time choosing one of the keys at random from her pocket. (All the keys look the same but she does not put a key back in her pocket once she has tried that key.) Conduct a simulation to estimate the probability it will take more than four tries to find the right key.

Exit Ticket

Your math teacher owns 10 ties and randomly chooses a tie to wear to work each school day (not much fashion sense). You notice he sometimes wears the same tie more than once during the week. You wonder if this is likely to happen often, so you decide you would like to find an estimate for the probability he wears the same tie more than once in a five-day workweek. To find this estimate, you design and conduct a simulation.

1. Describe the simple event.

Answer: Randomly choosing a tie.

2. Describe a model that would be appropriate to use for the simple event.

Answer: Number the ties from 1 to 10

Exit Ticket Cont.

3. Describe a trial and what you would record for each trial.

Answer: *Randomly choose five numbers between 1 and 10. The numbers could be chosen using a 10-sided die, randomly selecting numbers from 1 to 10 from a hat or bag or using the randint(1,10,5) function of the TI-84 graphing calculator or another rolling die simulator.*

Record whether or not a number is repeated.

4. Using the results below, what is an estimate for the probability he wears the same tie more than once in a five-day workweek?

Answer: *An estimate for the probability that he wears the same tie more than once in a five-day workweek is 17/28 = 0.61.*

Table 12.6: Results for the Simulation

Trial Number	Wears Same Tie More Than Once (Y/N)	Trial Number	Wears Same Tie More Than Once (Y/N)
1	Y	15	N
2	N	16	Y
3	N	17	Y
4	N	18	Y
5	N	19	Y
6	Y	20	Y
7	Y	21	Y
8	N	22	N
9	N	23	Y
10	Y	24	N
11	Y	25	Y
12	Y	26	N
13	Y	27	N
14	Y	28	Y

Further Explorations and Extension

Investigating the flu probability problem further.

The simulation gave an estimate for the probability of all six getting the flu. Using formal probability rules, find the exact probability of 2, 3, 4, 5, and 6 people getting the flu. Compare these answers to the simulated probability distribution developed in this lesson.

The probabilities can be calculated in the following manner:

Let X = the number of people infected

P(X=2) = 5/5 * 1/5 = 1/5 = 0.2

P(X=3) = 5/5 * 4/5 * 2/5 = 40/125 = 0.32

P(X=4) = 5/5 * 4/5* 3/5* 3/5 = 180/625 = 0.288

P(X=5) = 5/5 * 4/5* 3/5* 2/5 *4/5 = 480/3125= 0.1536

P(X=6) = 5/5 * 4/5 * 3/5* 2/5 * 1/5 = 120/3125 = 0.0384

Explanation for P(X=3)

5/5 = the probability Person 1 picks another person

4/5 = the probability Person 2 picks another person other than Person 1

2/5 = the probability the third person picks Person 1 or Person 2, which stops the flu.

Answer: Table 12.7 Probability Distribution

Table 12.7

X	P(X)
2	0.2
3	0.32
4	0.288
5	0.1536
6	0.0384
Total	1.0

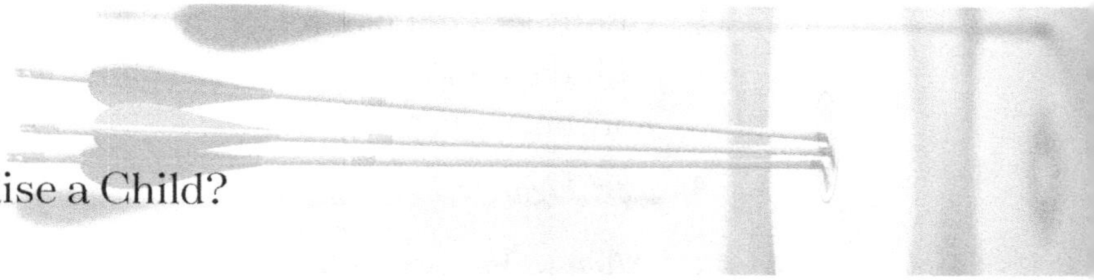

Investigation 13

What Is the Expected Cost to Raise a Child?

Expected Value

Overview

This session begins by introducing the definition of expected value of a random variable. Probability distributions are used to describe and model the behavior of a random variable. The expected value of the probability distribution is calculated and interpreted as the mean of the probability distribution.

This investigation follows the four components of statistical problem solving put forth in the *Guidelines for Assessment and Instruction in Statistics Education (GAISE) Report*. The four components are formulate a statistical question, design and implement a plan to collect data, analyze the data by measures and graphs, and interpret the results in the context of the original question. This is a GAISE Level B activity.

This activity is based on lessons from *Probability Models*, published by the American Statistical Association (original copyright by Dale Seymour Publications 1999). *Probability Models* is a module in the ASA Data-Driven Mathematics Project. It is available as a free download *www.amstat.org*.

Instructional Plan

Brief Overview

» Introduce the concept of *expected value*, the notation and connection to the mean of a probability distribution.

» Read and discuss the scenario pertaining to the Nielsen ratings.

» Formulate the statistical question: "If a US family is randomly selected, how much would you expect the cost to be for raising all the children in the family for one year?"

» Find the expected value of the given probability distribution of the number of children under 18 in a family.

» *Optional:* Application of expected value

Introducing Expected Value of a Random Variable

Handout Student Worksheet 13.1 Introducing Expected Value.

During the 2018 Major League Baseball playoffs, one team played in 10 games before losing the division series. In the 10 games, they scored one run in two games, two runs in one game, four runs in four games, and six runs in three games.

1. Complete the frequency table below.

Answer:

Number of Runs	Frequency
1	2
2	1
3	0
4	4
5	0
6	3

2. Construct a dot plot for the runs scored with the variable "runs scored" on the horizontal axis.

Answer: Figure 13.1

3. What is the mean number of runs scored per game for the team? Explain how you found the answer and interpret the mean.

Answer: 3.8 runs.
[1(2) + 2(1) + 4(4) + 6(3)] / 10 = 3.8

Suppose you knew the team scored one run in 20% of the games, two runs in 10% of the games, four runs in 40% of the games, and six runs in 30% of the games, but you didn't know the number of games played.

4. Construct a histogram of the number of runs scored with the vertical axis as relative frequency. How does this graph compare with the dot plot in problem 2?

Possible answer: *The graph in Figure 13.2 is similar, except the vertical axis represents the fraction of games played, rather than number of games played.*

5. Calculate the mean number of runs scored. Explain how you found the answer.

Learning Goal

Understand how to compute and interpret the expected value of a random variable as the mean of the probability distribution.

Mathematical Practices Through a Statistical Lens

MP4. Model with Mathematics

Statistically proficient students can apply mathematics to help answer statistical questions arising in everyday life, society, and the workplace.

Materials

Student worksheets are available at *www.statisticsteacher.org/statistics-teacher-publications/focus.*

- Student Worksheet 13.1 Introducing Expected Value
- Student Worksheet 13.2 Applying Expected Value
- Optional: Student Worksheet 13.3 Application of Expected Value
- Optional: Census Bureau website: *www.census.gov*
- Exit Ticket

Estimated Time

One to two 50-minute class periods

Pre-Knowledge

Students should already be able to find and interpret the weighted mean of a distribution.

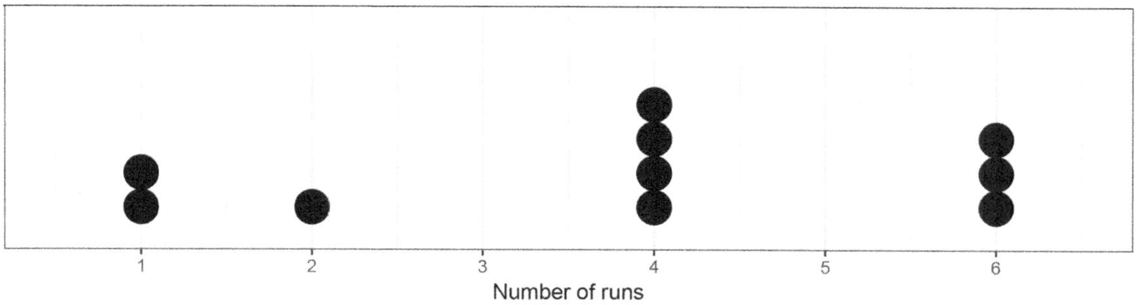

Figure 13.1: Dot plot for the runs scored

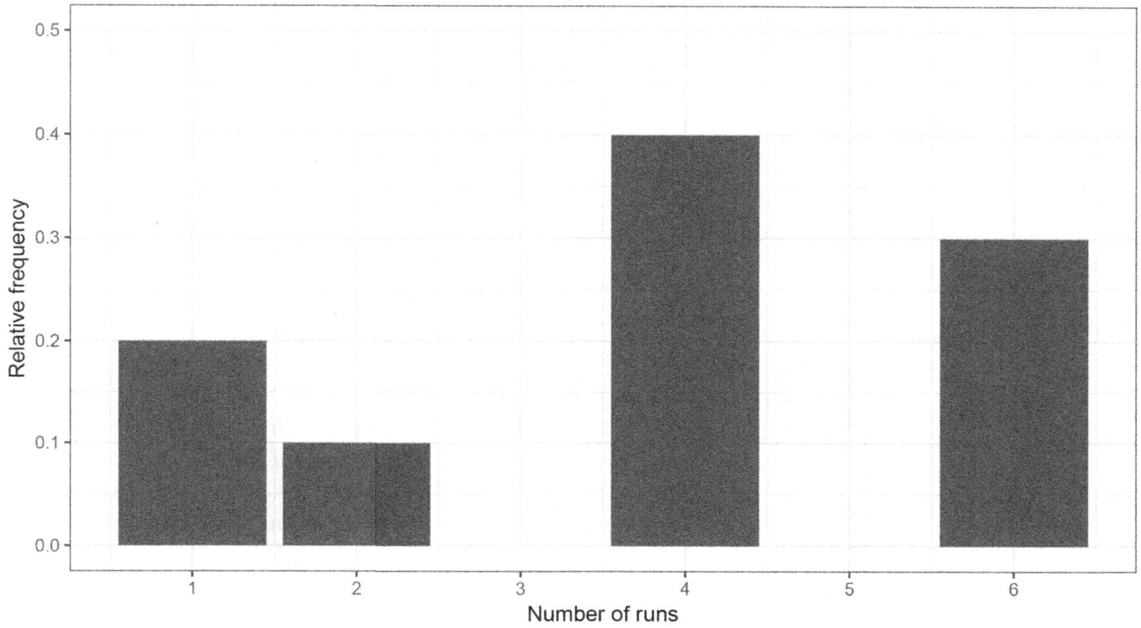

Figure 13.2: Histogram of the number of runs scored

Answer: *3.8 runs.*
1(.20) + 2(.10) + 4(.40) + 6(.30) = 3.8

The team is expected to perform about the same at the start of the next season as they did in the playoffs. That is, the probability of scoring one run in a game is about 20%, scoring two runs in a game about 10%, scoring four runs in a game about 40%, and scoring six runs in a game about 30%.

A mean calculated from a probability distribution—an anticipated distribution of outcomes—is called an expected value.

Let the variable X = the number of runs scored by the team. The variable X is called a random variable.

The probability distribution for a random variable can be displayed in a two-column table, like Table 13.1, for the variable X—the number of runs scored. The symbol P(X) represents the probability of the team scoring X runs in a randomly selected game.

A commonly used symbol for the expected value of X is E(X).

6. Write a symbolic expression for the expected value of X. Recall how you

Table 13.1

X	P(X)
1	0.2
2	0.1
3	0
4	0.4
5	0
6	0.3

Table 13.2

X	P(X)
x_1	p_1
x_2	p_2
x_3	p_3
x_k	p_k

Table 13.3

Number of Motor Vehicles	Relative Frequency (Rounded to 2 Decimal Places)
0	0.09
1	0.34
2	0.37
3	0.14
4	0.06

explained the method used to find the mean of the distribution.

Note: This may be a difficult question, and students may need some scaffolding. One suggestion is to set up a table showing x_1, x_2, x_3, ... x_k under a column labeled X and p_1, p_2, p_3, ... p_k under a column labeled P(x). Then, remind students how they would find the mean.

Possible Answer: *Table 13.2*

Expected value of X could be written $E(X) = x_1p_1 + x_2p_2 + x_3p_3 + \dots + x_kp_k$

or

$$E(X) = \sum_{i=1}^{k} x_i p_i$$

Application of Expected Value (Optional)

Hand out Student Worksheet 13.2 Applying Expected Value.

Table 13.3 shows the distribution of the number of motor vehicles per US household. A "household" is defined by the US Census Bureau as all persons occupying a housing unit such as a house, an apartment or other group of rooms, or a single room. *Source: www.census.gov*

1. The sum of the relative frequencies is 1.00. Does that mean no US household has more than four cars?

Possible Answer: *A household could have more than four cars, but the relative frequency of such households would round to 0.00.*

Suppose the Department of Energy is planning to select a random sample of households in the US to conduct a survey about reformulated gasoline.

Define a random variable M to be the number of motor vehicles in a randomly selected US household.

2. Find and interpret E(M), the expected value of M.

Answer: *We expect about 1.74 cars per US household.* $0(0.09) + 1(0.34) + 2(0.37) + 3(0.14) + 4(0.06) = 1.74$

3. If the Department of Energy randomly selected 1000 households, how many motor vehicles would we expect these households to have?

Answer: *1.74(1000) = 1740 cars*

Scenario

Have your students read the scenario and answer questions 4 to 11.

Have you and your family ever taken part in a TV or radio rating survey? Maybe you were asked what TV shows you watched or what radio station you listened to on a regular basis. Have you heard of the Nielsen rating?

The Nielsen Corporation is a global marketing research firm. This company was founded in 1923 in Chicago by Arthur C. Nielsen Sr. to give marketers reliable and objective information about the impact of marketing and sales programs. One of Nielsen's best known creations is the Nielsen ratings, a system that measures how many people are watching different TV shows or listening to different radio stations. Nielsen uses statistical sampling to randomly select a representative sample of about 5000 households who agree to be part of the rating estimates. To find out what shows people are watching, meters are installed on all the TV sets in the household. These meters keep track of what TVs are on at any given time and what show the TV set is tuned to. *Source: https://en.wikipedia.org/wiki/Nielsen_Corporation*

4. Why might Nielsen ratings be important information for a TV or radio station?

Possible answer: *Higher ratings mean the station can set higher advertising rates.*

Imagine the television network Nick Jr. is interested in what TV shows people under the age of 18 watch. Network executives could ask the Nielsen Corporation for help determining the most-watched children's TV shows.

The Nielsen Corporation plans on selecting a random sample of families across the US and is particularly interested in families with children under the age of 18. Prior to conducting the survey, researchers at Nielsen find data on the number of children in US families on the US Census Bureau website. According to the US Census Bureau, the number of children under 18 years of age per family in 2010 has a distribution shown in Table 13.4. A "family" is defined as a group of two or more persons related by birth, marriage, or adoption, residing together in a household.

Table 13.4

Number of Children Under 18 in a Family	Percent (Rounded to 1 Decimal Place)
0	55.3
1	19.2
2	16.4
3	8.1
4	1.0

Refer to the distribution of number of children under 18 and answer the following questions.

5. Why do you think the percentage of families with 0 children is so high?

Possible Answer: A large number of families have children older than 18 or no children.

6. The sum of the percentages equals 100%. Does that mean no U.S. families have more than four children?

Possible Answer: *A family could have more than four children, but the relative frequency of such households is tiny and would round to 0.0.*

7. Construct a histogram of the number of children under 18 in US families.

Possible Answer: *Figure 13.3*

8. Find the mean and interpret the mean. Locate the mean on the horizontal scale of the histogram. Is the mean in the center of the graph? Why or why not?

Answer: *The mean is 0.803. The mean is not in the center of the graph. It is much closer to 0 because 0 has such a high frequency of occurrence.*

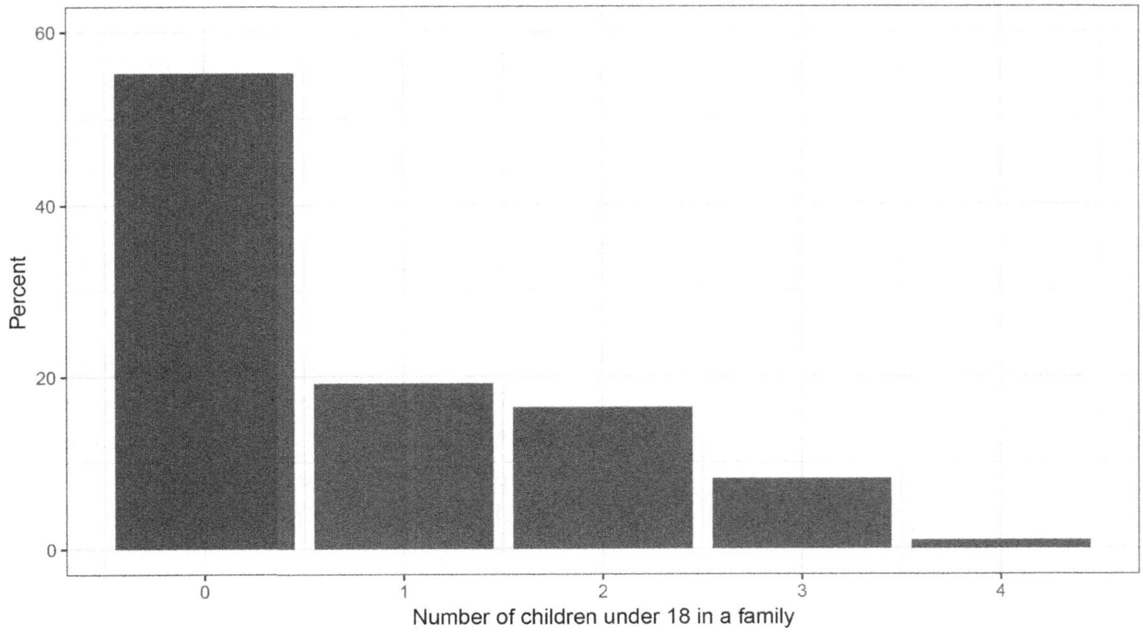

Figure 13.3: Histogram of the number of children under 18 in US families

The A.C. Nielsen Company randomly selects families for use in estimating the ratings of TV shows. Let the variable N represent the number of children under 18 in a randomly selected US family.

9. If A.C. Nielsen randomly selected a family, what is the expected value of N, the number of children under 18 in a randomly selected family?

***Answer:** 0.803 children per family*

10. How many children in all would we expect to see in a random sample of 2500 families?

***Answer:** 2500(0.803) is approximately 2008 children.*

11. If Nielsen wants the opinion from at least 1000 children, how many families should be in their random sample?

***Answer:** 1000/0.803 is approximately 1245 families.*

Formulate a Statistical Question

Suppose families can expect to spend around $13,000 a year to raise a child. Housing, food, child care, clothing, health care, and transportation are some of the expenses.

We want to study a probability distribution for a new random variable C, the cost to raise a child for one year.

Ask your students to consider the statistical question: "If a US family is randomly selected, how much would you expect the cost to be for raising all the children in the family for one year?"

Collect Appropriate Data

12. Using the data from Nielson pertaining to the number of children per family and the cost of raising a child per year, complete the probability distribution (Table 13.5). The first column should contain all the possibilities for C—the cost to raise children in a family for one year.

Answer: Table 13.5

Analyze the Data

13. If a US family is randomly selected, how much would you expect the cost to be for raising all the children in the family?

Answer: 0(0.553) + 13000(0.192) + 26000(0.164) + 39000(0.081) + 52000(0.01) = $10439

Interpret the Results in the Context of the Original Question

14. What was the expected value of N, number of children under 18 in a randomly selected US family?

Answer: 0.803 children per family

15. How is the expected value of C related to the expected value of N?

*Answer: E(C) = 13000*E(N)*

16. Could you have found the expected value of C without building the probability distribution?

Answer: Yes, multiply the cost to raise a child times expected value of N.

Optional: If students need another example, hand out Student Worksheet 13.3 Application of Expected Value.

Scenario

The high-school band is selling raffle tickets to raise money for new uniforms. The winner of the random drawing will receive a necklace designed and made by one of the band parents. The raffle tickets cost $1, and the necklace has a value of $100. The band sells 200 tickets.

Let G represent the amount gained if you buy one ticket.

There are two possible outcomes—you win or lose. If you lose, you have lost your $1, which can be a gain of -1. If you win, your gain would be 100 minus the 1 for the ticket, or a gain of $99.

Table 13.5

C	P(C)
0	0.553
13000	0.192
26000	0.164
39000	0.081
52000	0.01

Table 13.6

Gain/Loss	Probability of Gain/Loss
-$1	199/200
$99	1/200

Table 13.7

Gain/Loss	Probability of Gain/Loss
-$10	190/200
$90	10/200

1. If a person buys one raffle ticket, find the probability distribution for the gain/loss.

Answer: Table 13.6

2. Find the expected gain/loss for a player who buys one raffle ticket.

Answer: Expected value is -0.50, or lose 50¢ for every $1 raffle ticket purchased.

3. What would the expected gain/loss be if a person bought 10 tickets?

Answer: Table 13.7. Expected value is lose $5.

4. What would the expected gain/loss be if a person bought 100 tickets? Can you find the answer without creating a probability distribution?

Answer: Lose $50.

Additional Ideas

Using the Census Bureau website, *www.census.gov*, find the data on the number of TV sets per US household. Using the data, have the students answer the question: "If a US household is randomly selected, what is the expected value of the number of TV sets per US household?"

Exit Ticket

A mobile phone company offers an optional protection that will pay for repairs if the phone is damaged in an accident. The plan costs $50. The retailer has determined the typical cost to repair a broken phone is $150. Let R be the number of repairs a randomly chosen customer will use under this plan. Following is the probability distribution for R.

R	P(R)
0	0.80
1	0.17
2	0.02
3	0.01

1. What is the expected value of R, the number of repairs needed by a randomly selected customer?

Answer: *0.24 repairs*

2. Let C represent the amount it will cost the phone company in repairs for a randomly selected customer. Find the expected value of C.

Answer: 0.24(150), or $36

3. What is the expected amount of profit the company will make from a randomly selected customer?

Answer: 50–36 = 14. The company is expected to make $14 for each randomly selected customer.

Further Explorations and Extensions

Have students find and interpret the standard deviation of a probability distribution.

Refer to the probability distribution for the random variable N—the number of children under 18 in a randomly selected US family.

N	P(N)
0	0.553
1	0.192
2	0.164
3	0.081
4	0.01

Possible answer: *The standard deviation of 1.045 is the number of children Nielsen would expect to typically vary per randomly selected family from the expected value of 0.8.*

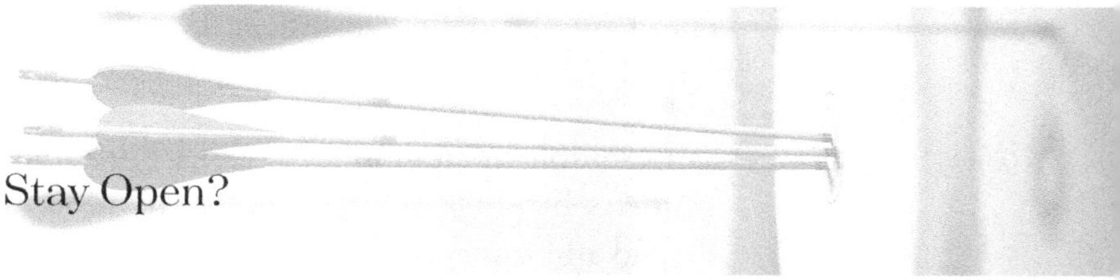

Investigation 14

How Long Do the Subway Doors Stay Open?

Normal Distribution

Overview

This investigation introduces the Normal distribution as a possible model to describe a sample of times subway doors stay open. The empirical rule is developed and used to help decide if the Normal distribution can be used to model a sample of times. Students will also use the empirical rule to decide what typical door-open times are.

This investigation follows the four components of statistical problem solving put forth in the *Guidelines for Assessment and Instruction in Statistics Education (GAISE) Report.* The four components are formulate a question, design and implement a plan to collect data, analyze the data by measures and graphs, and interpret the results in the context of the original question. This is a GAISE Level C activity.

Instructional Plan

Brief Overview

- Introduce the empirical rule.
- Develop a statistical question pertaining to the length of time subway doors stay open.
- Decide if the Normal distribution can be used to model the distribution of times subway doors stay open.

Introducing the Normal Distribution

Remind your students that in many of the earlier investigations, we encountered different shapes of distributions. Some were

skewed—like the memory test times (Investigation 3)—and others were mound shaped and symmetric—like the length of sample baseball games in 1987 (Investigation 2). The mound-shaped and symmetric distributions are common and key in the study of statistics. Some of these distributions are often referred to as the Normal curve or Normal distribution. Data distributions that are mound shaped and symmetric are often modeled with the Normal curve or Normal distribution.

There are many examples of the application of the Normal distribution. Healthy biological populations such as the heights and weights of animals follow a Normal distribution.

Other examples of Normal distributions include standardized test scores, a person's blood pressure, the weight of packages of cookies, and IQ.

Hand out Student Worksheet 14.1 Normal Distribution.

The example in Figure 14.1 is a distribution of women's heights. The distribution is mound shaped and somewhat symmetric with a mean of 64.6 inches and standard deviation of 2.75 inches. So, we might say the distribution appears approximately Normal.

Another example (Figure 14.2) shows the achievement scores for 200 students at a high school. The distribution is mound shaped and somewhat symmetric with a mean of 75.7 and a standard deviation of 6.1.

Each Normal curve is unique based on the mean and standard deviation, but all have the same shape and properties.

Learning Goals

- Describe a distribution as mound shaped and approaching a Normal distribution.
- Estimate population percentages using the empirical rule.

Mathematical Practices Through a Statistical Lens

MP4. Model with Mathematics

Statistical models build on mathematical models by including descriptions of the variability present in the data.

Materials

Student worksheets are available at *www.statisticsteacher.org/statistics-teacher-publications/focus.*

- Statistical software or app capable of finding summary statistics and constructing a histogram.
- Student Worksheet 14.1 Normal Distribution
- Student Worksheet 14.2 Scenario
- Student Worksheet 14.3 Analyze the Data
- Exit Ticket

Estimated Time

Two 50-minute class periods. One period to introduce the Normal distribution and empirical rule. Another period to determine if a distribution is approximately normal.

Pre-Knowledge

Students should already be able to:

- Use technology to find the mean and standard deviation
- Use technology to construct a dot plot and histogram

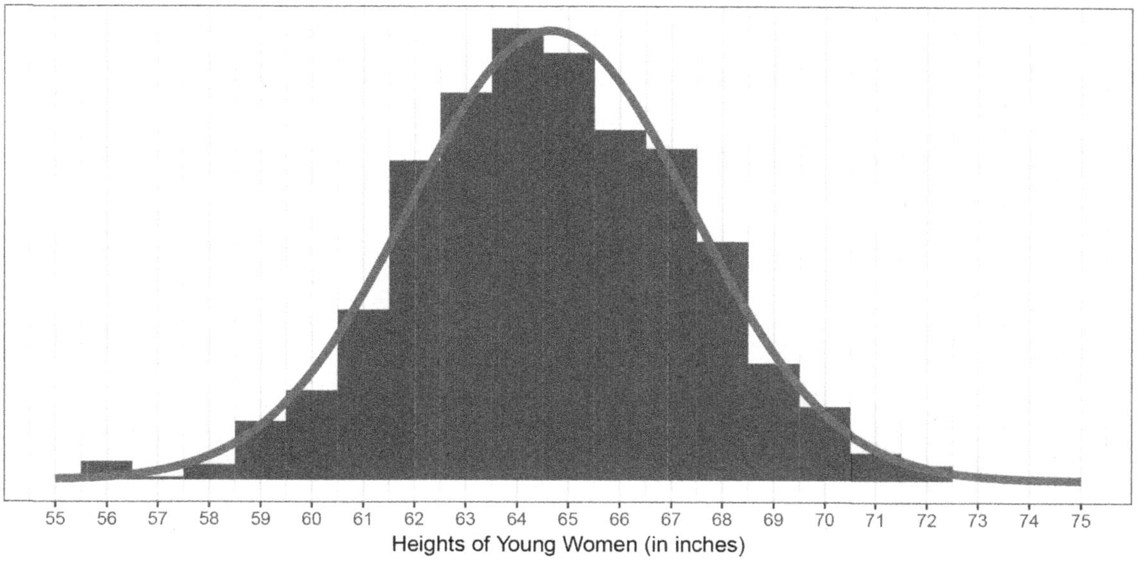

Figure 14.1: Distribution of women's heights

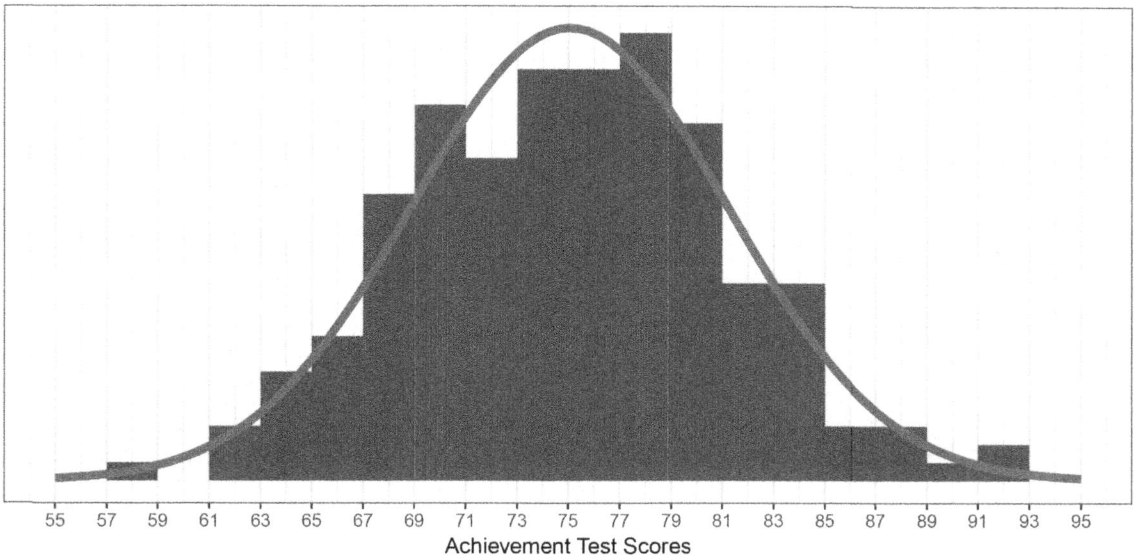

Figure 14.2: Distribution of achievement scores for 200 students at a high school

The proportion of data within one and two standard deviations of the mean is the same for all Normal curves. These proportions form the empirical rule.

The empirical rule states that for a Normal distribution, nearly all the data will fall within three standard deviations of the mean.

The empirical rule can be broken down into three parts, as shown in Figure 14.3:

» Approximately 68% of data falls within one standard deviation of the mean.

» Approximately 95% of data falls within two standard deviations of the mean.

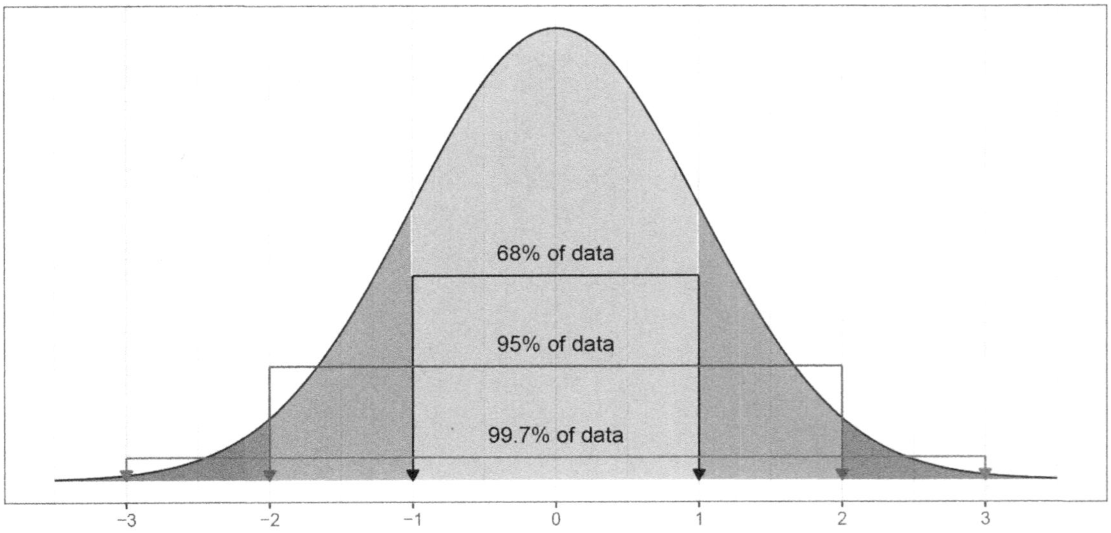

Figure 14.3: The three parts of the empirical rule

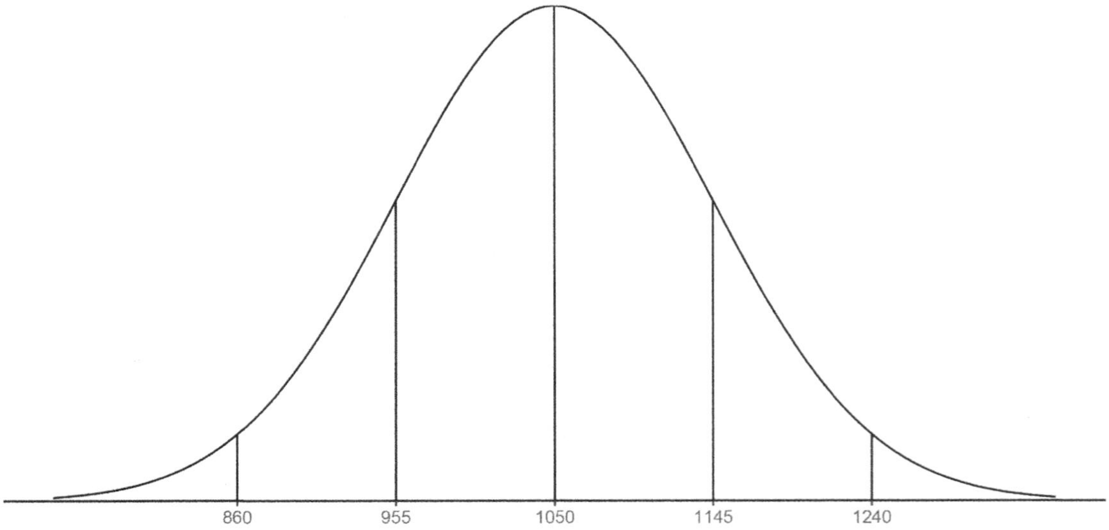

Figure 14.4: Mean and values for one and two standard deviations from the mean

» Approximately 99.7% of data falls within three standard deviations of the mean.

Using the Empirical Rule

Suppose the life of a certain brand of light bulbs can be modeled with the Normal distribution with a mean of 1050 hours and a standard deviation of 95 hours.

1. On the Normal curve in Figure 14.4, add the mean and values for one and two standard deviations from the mean.

Answer: Figure 14.4

2. What proportion of bulbs lasts between 955 and 1145 hours?

Answer: Approximately 68%

3. What proportion of bulbs lasts between 860 and 1240 hours?

Answer: Approximately 95%

4. What proportion of bulbs lasts less than 860 hours?

Answer: *Approximately 2.5%*

5. What proportion of bulbs lasts between 955 and 1050 hours?

Answer: *Approximately 34%*

6. What proportion of bulbs lasts between 1145 and 1240 hours?

Answer: *Approximately 13.5%*

Modeling with the Normal Curve

Hand out Student Worksheet 14.2 Scenario. Give your students time to read the scenario. After they have read the scenario, ask if they have additional questions about the subway door operation.

Scenario

If you live in a large city, it is likely you will encounter a subway or similar mass transportation system. People who live in Washington DC frequently use the Metro. They use the BART in San Francisco, the "L" in Chicago, the MBTA in Boston, the DART in Dallas, and the monorail at Disney World. Mass transportation, however, requires people to make decisions that influence when and where they make their connections. How much time is needed to get to my destination? When should I leave? How long will the doors stay open, or how long will the train wait before leaving? Students in a New York City high school indicated that to get to school on time, they must consider not only when the best time to catch a subway connection is, but also what the chance is a door on the subway will close before they have a chance to get on the train. The students indicated to their teacher that their tardiness to school is often a result of the doors closing before they have a chance to get on board.

In planning the route to school, the following questions were considered by the students:

» When will the subway connection arrive at our location?

» How many people will generally connect at this location?

» What is the expected time the doors will stay open to catch the subway?

Several students indicated that they frequently missed their connection because the doors did not stay open long enough, while other students indicated it rarely happened to them. The students asked a number of questions related to the subway door operation.

» How long do the doors of a subway stay open?

» Are the doors opened and closed automatically or manually?

» Do the doors stay open approximately the same amount of time throughout the day?

» Does the number of people in the subway possibly influence the length of time?

After exploring several factors that might influence the opening and closing of the doors, the students at this school decided to investigate some of these questions through a statistical study. They were particularly interested in the urban myth that the doors on the New York City subway stay open for 30 seconds since they based their decisions of when to catch a subway based on that time.

Formulate a Statistical Question

Point out to your students that the doors on the New York City subway cars are supposed to stay open for 30 seconds according to the urban myth in the scenario. We want to determine if

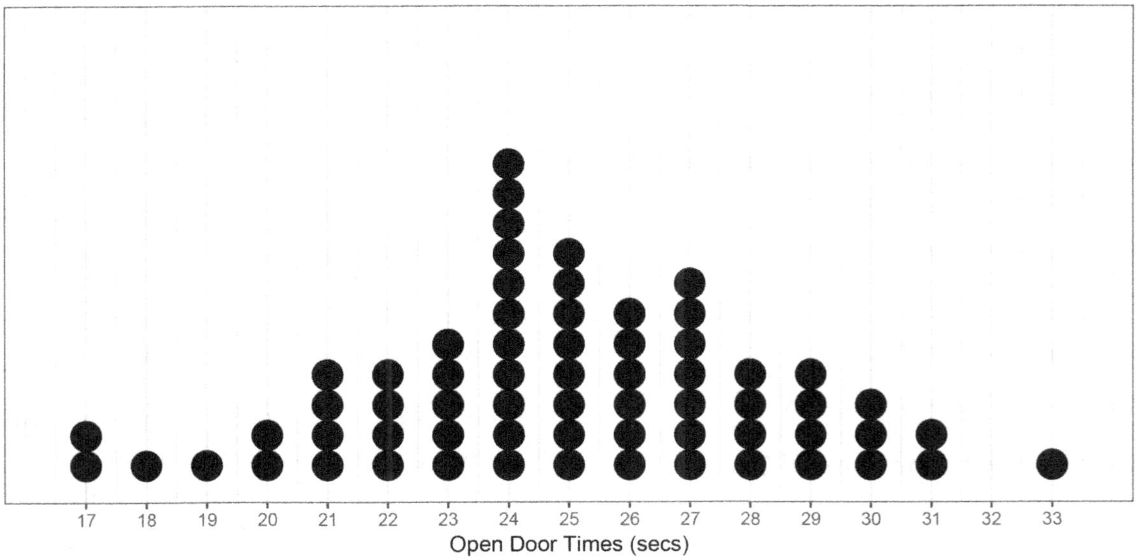

Figure 14.5: Dot plot of the door open times

this myth has validity. Ask your students to consider the statistical questions: "How can we model how long the doors on the F train stay open?" "Is a Normal distribution an appropriate model for the length of time the doors on the F train on the New York City subway route stay open?"

Collection of Data

Ask your students how they would collect times that the doors on the F train stay open. What problems might they encounter in collecting the data?

Answers will vary. They might suggest collecting times at different times of the day, rather than just over the lunch periods, or at different subway stops.

Hand out Student Worksheet 14.3 Analyze the Data.

Discuss with your students how a group of high-school students in Brooklyn, NY, collected data to help answer the statistical questions.

Over a period of 18 school days, a group of students in Brooklyn, New York, recorded how long the doors on the F train stayed open at the subway stop near their school. They collected 65 lengths of time, to the nearest second. Each measurement was taken at approximately the same times each day, usually during three lunch periods.

Below are the data collected:

31 17 19 20 33 29 25 25 26 17 18 29 22 24
24 26 30 27 21 23 28 25 24 21 20 31 28 27
21 24 28 29 30 21 24 27 25 24 23 22 30 26
26 25 25 24 29 24 25 27 27 24 26 22 23 27
22 24 26 28 23 24 25 23 27

Analysis of the Data

Ask your students to answer questions 1 to 4.

1. Construct a dot plot of the door-open times.

Answer: Figure 14.5

2. Describe the distribution of the subway door-open times. Include in your description the shape, an estimate for the mean, and an estimate for the standard deviation.

***Possible answer:** The distribution is mound shaped with a mean of about 25 sec. and a standard deviation of about 3 sec.*

3. Interpret the mean and standard deviation in this context.

Possible answer: *The mean of 25 seconds is the balance point of the lengths of time deviations on a dot plot. The standard deviation of 3 sec. is the typical length of time a door stays open as measured from the mean.*

4. Using the distribution of subway door-open times, how would you answer the question, "What is the length of time the doors on the F train on the New York City subway route typically stay open?"

Possible answer: *The distribution of subway door-open times centers around 25 seconds. It appears a typical length of time the subway doors are open is approximately 25 seconds. Most of the times were between 21 and 30 seconds. There were only six recorded times when the subway doors stayed open 30 or more seconds.*

After discussing the questions, ask your students to answer questions 5 to 9.

5. To help further understand the distribution of times, complete the frequency chart.

Answer: *Table 14.1: Length of Time Subway Doors Were Open on the F Train*

Table 14.1

Lengths of Time (sec)	Frequency	Lengths of Time (sec)	Frequency
17	**2**	25	**8**
18	**1**	26	**6**
19	**1**	27	**7**
20	**2**	28	**4**
21	**4**	29	**4**
22	**4**	30	**3**
23	**5**	31	**2**
24	**11**	32	**0**
		33	**1**
		Total	**65**

6. What percent of door-open times were:
   a. Less than or equal to 24 seconds?
      Answer: *30/65 = 0.46, or 46%*
   b. More than or equal to 30 seconds?
      Answer: *6/65 = 0.17, or 17%*
   c. Between 22 and 28, inclusive?
      Answer: *45/65 = 0.692, or 69.2%*
7. Convert the frequencies to relative frequencies and record in Table 14.2.
8. Use technology to construct a histogram (bin width of 1 sec.) of the subway door-open times with relative frequency as the

Table 14.2

Lengths of Time (sec.)	Frequency	Relative Frequency	Lengths of Time (sec.)	Frequency	Relative Frequency
17	2	**0.031**	25	8	**0.123**
18	1	**0.015**	26	6	**0.092**
19	1	**0.015**	27	7	**0.108**
20	2	**0.031**	28	4	**0.062**
21	4	**0.062**	29	4	**0.062**
22	4	**0.062**	30	3	**0.046**
23	5	**0.077**	31	2	**0.031**
24	11	**0.169**	32	0	**0**
			33	1	**0.015**
				Total	**1.001**

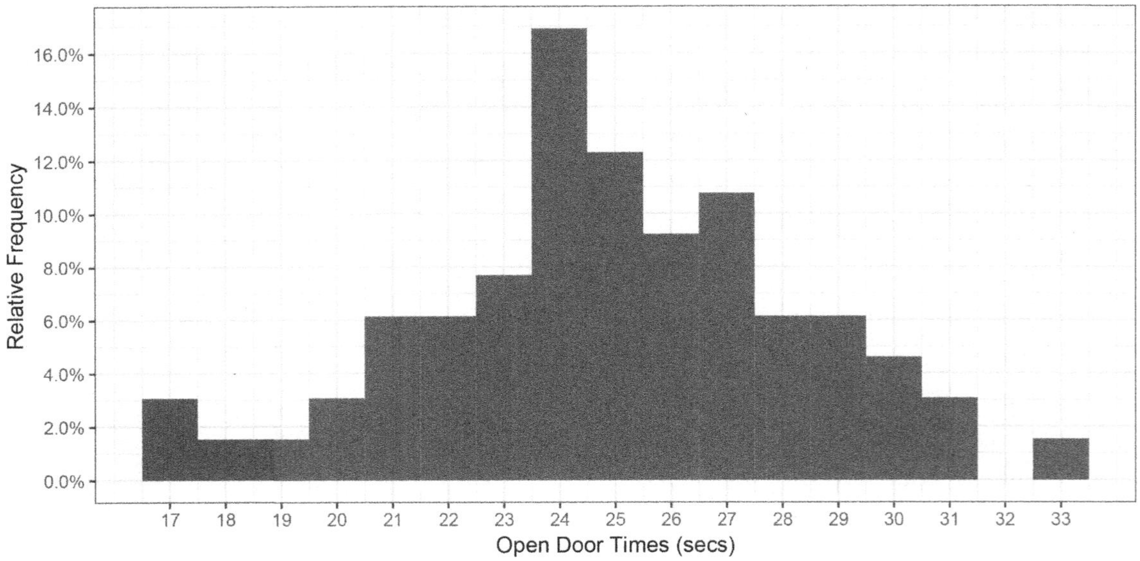

Figure 14.6: A histogram of the subway door open times with relative frequency as the y-axis.

y-axis. Use technology to find the mean and standard deviation.

Note: Students may need help finding the mean and standard deviation with grouped data.

Possible answer: *Figure 14.6. The distribution of subway open door times has a mean and standard deviation of 24.9 sec. and 3.4 sec., respectively.*

9. What are the similarities and differences between the two graphs (dot plot and histogram)?

Possible answer: *Both graphs of the distribution of subway door-open times have the same shape—approximately mound shaped—same center around 25 seconds, and same standard deviation. The only difference is the scale on the y-axis.*

After discussing questions 5 to 9, ask your students to complete Question 10. Have a brief discussion about their responses.

10. Do you think a Normal distribution is an appropriate model for the length of time the doors on the F train on the New York City subway route stay open?

Answers: *Answers will vary, but many students will say the distribution appears to be approximately Normal.*

Explain to your students that we compare some of the properties of a Normal distribution to the distribution of times to further investigate whether the Normal distribution is a good model for the times the subway door stays open.

To do that, let's assume a Normal distribution is an appropriate model with a mean of 24.9 sec. and a standard deviation of 3.4 sec.

As a class, work through questions 11 to 16.

11. On the Normal curve below, locate the mean and the times within one and two standard deviations of the mean by drawing a vertical line through these points.

Answer: *Figure 14.7*

12. Using the empirical rule, what proportion of the data is within one standard deviation of the mean? Show this on the Normal curve.

***Answer:** Figure 14.8; approximately 68%*

13. Using the empirical rule, what proportion of the data is within two standard deviations of the mean? Show this on the Normal curve.

***Answer:** Figure 14.9; approximately 95%*

14. Using the data table of the relative frequencies, approximately what percent of the open-door times are within one standard deviation of the mean?

***Answer:** 24.9 − 3.4 = 21.5, 24.9 + 3.4 = 28.3. The sum of the relative frequencies from 22 seconds to about 28 seconds is approximately 0.062+0.077+0.169+0.123+0.092+0.108+0.062 = 0.692, or 69.2%.*

15. Using the data table of the relative frequencies, approximately what percent of the open-door times are within two standard deviations of the mean?

***Answer:** 24.9 − 2*3.4 = 18.1, 24.9 + 2*3.4 = 31.7. The sum of the relative frequencies from 18 to 32 is approximately 0.693+.015+.015+.031+.062+.062+.046+.031= 0.955, or 95.5%.*

16. How do the proportions you found in questions 14 and 15 compare with the empirical rule percentages based on the Normal curve as a model for the times the subway doors are open?

***Possible answer:** The percentages are very close to the theoretical percentages.*

Interpret the Results in the Context of the Original Question

Give students time to answer questions 17 to 19.

17. Do you think a Normal distribution is an appropriate model for the length of time the doors on the F train on the New York City subway route stay open?

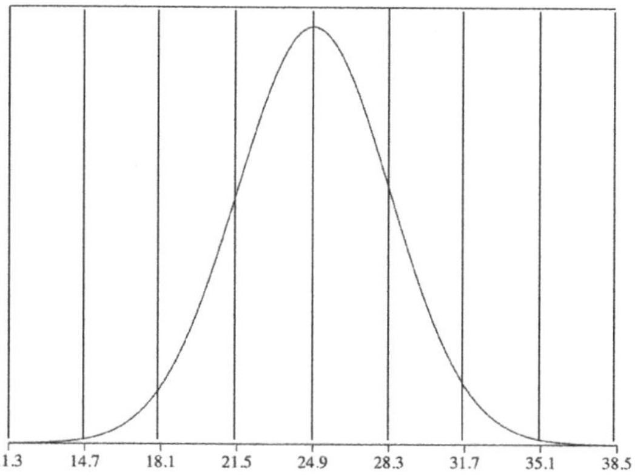

Figure 14.7: The mean and times within one and two standard deviations of the mean

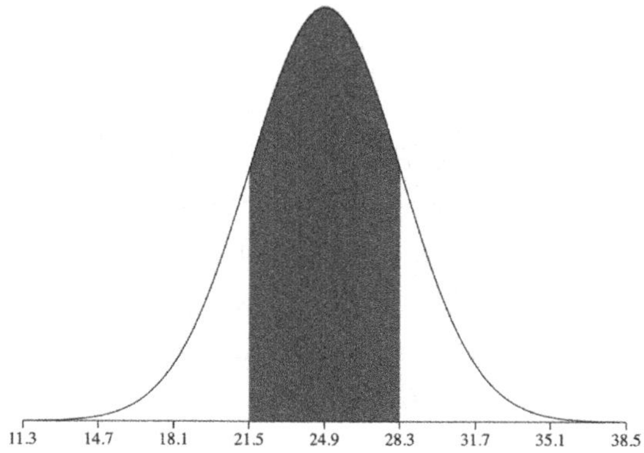

Figure 14.8: The proportion of data within one standard deviation of the mean

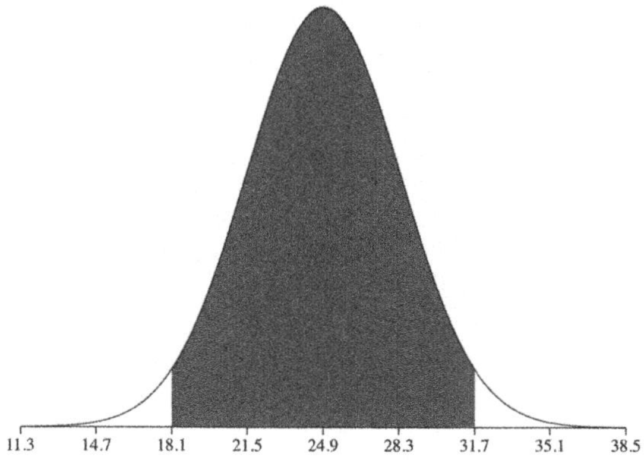

Figure 14.9: The proportion of data within two standard deviations of the mean

***Answer:** The distribution is mound shaped and symmetric, and the percent of data within one and two standard deviations of the mean approximately equal the percentages found in the empirical rule.*

18. Use the model (Normal curve) you have created and decide if the urban myth that the subway doors typically stay open for 30 seconds is true.

Possible answer: *It seems unlikely the doors stay open for 30 seconds. Out of the 65 times collected, the doors stayed open 30 or more seconds only six times, or less than 10% of the time. It appears the length of time the doors stay open is approximately a Normal distribution with a mean of about 25 seconds and a standard deviation of about 3.5 seconds. Thirty seconds is well outside what we would expect.*

19. Another method to determine whether there are outliers is to use the standard deviation. If the distribution is mound shaped (approximately Normal), then any data point more that two standard deviations from the mean can be considered an outlier. Using this definition, are there any outliers in the subway door-open times? Is the urban myth of the doors staying open 30 seconds an outlier?

Possible answer: *Any value less than 18.1 seconds or three values at 18, 17, and 17. Any value greater than 31.7 or one value at 33. 30 seconds is not an outlier, but it is still outside the typical length of time the doors stay open.*

Additional Ideas

In lieu of using the data presented in this lesson, students could do the following:

» Collect data from the internet/newspaper on the price of a particular model of new car and explore the Normal distribution as a model for the distribution.

» Collect data from the internet/newspaper on the price of a particular model of used car and explore the Normal distribution as a model for the distribution.

» Collect the times to go through the lunch line in the school cafeteria and explore the Normal distribution as a model for the distribution.

» Collect data from the internet/newspaper on the batting averages for major league baseball players and explore the Normal distribution as a model for the distribution.

» Collect data from the internet on the length of Old Faithful's eruption times and explore the Normal distribution as a model for the distribution.

Exit Ticket

Alicia, a senior in high school, would like to find out how her SAT score compares with other seniors who took the SAT. She decided to investigate the question: "What is a typical score on the math portion of the SAT test?"

The College Board reported in 2016 that approximately 1.7 million high-school students took the SAT. The scores on the math portion of the test were approximately Normally distributed with a mean score of 513 and standard deviation of approximately 118.

1. Draw a Normal curve and mark the mean and one standard deviation and two standard deviations from the mean.

Answer: *Figure 14.10*

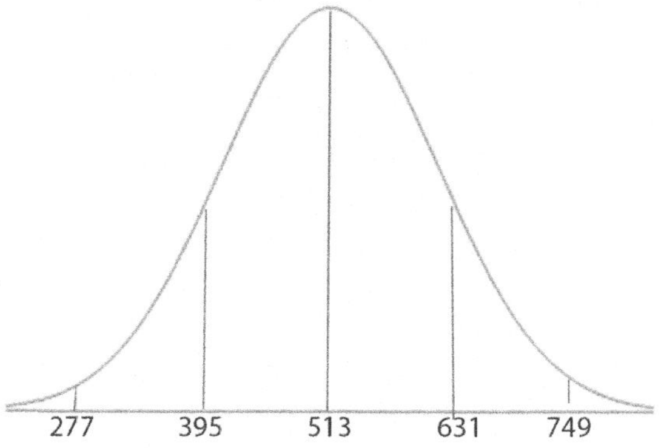

Figure 14.10: Mean with one and two standard deviations from the mean

2. What proportion of scores is within one standard deviation of the mean?

Answer: *Approximately 68%*

3. What proportion of SAT scores is between 277 and 749?

Answer: *Approximately 95%*

4. If Alicia had a score of 631, approximately what percent of the students did she do better than?

Answer: *Alicia scored higher than approximately 84% of the students who took the SAT test.*

Further Explorations and Extensions

The term "Normal curve" came from the study of the errors made in astronomical observations and other scientific observations. Abraham de Moivre, an 18th-century statistician and gambling consultant, was often asked to make lengthy computations, like the probability of observing at least 60 heads in 100 coin tosses. Because this calculation would be difficult, de Moivre observed as the number of tosses increased, the distribution of the number of heads became more and more mound shaped, symmetrical, and smoothly curved. If he could find a function for this curve, he could find the probability of getting 60 or more heads out of 100 coin flips with much less difficulty.

What we now call the Normal curve is what de Moivre discovered around 1733. The curve may also have been discovered separately by the astronomer and mathematician Pierre Laplace in 1786. A different derivation of the formula was presented in 1809 by Karl Friedrich Gauss, hence the Normal curve is often called the Gaussian curve.

Summarized from *https://onlinestatbook.com/2/normal_distribution/history_normal.html* and *https://sydney.edu.au/stuserv/documents/maths_learning_centre/normal2010web.pdf*

The Normal curve is mound shaped and symmetrical. One form of the equation is:

$$f(x) = \frac{1}{\sigma\sqrt{2\pi}} e^{-\frac{(x-\mu)^2}{2\sigma^2}}$$

μ is the population mean

σ is the population standard deviation

π is the constant pi

e is the constant, Euler's number, 2.718281828 …

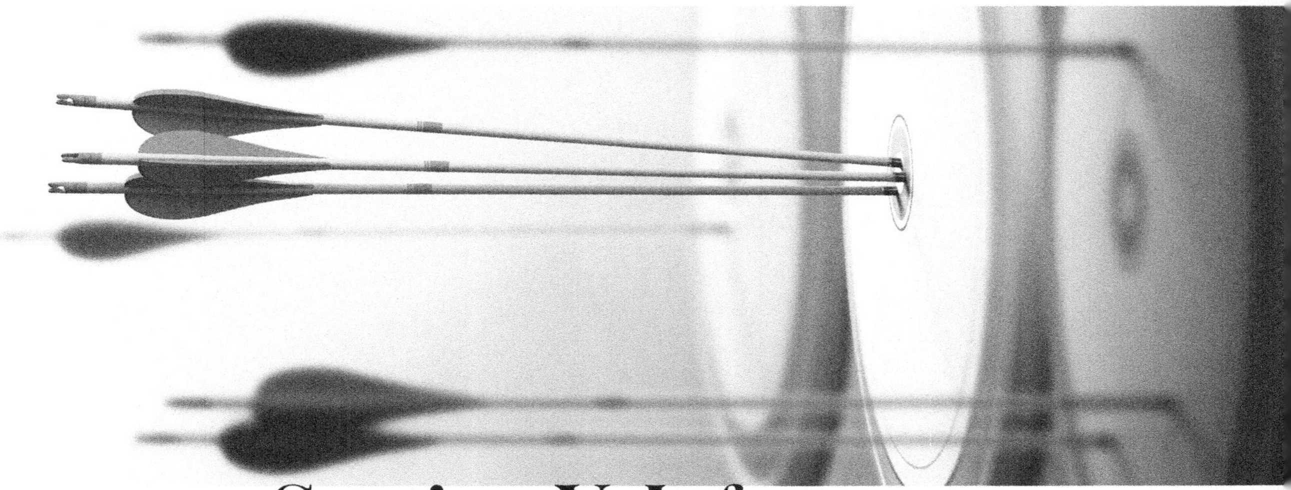

Section V: Inference

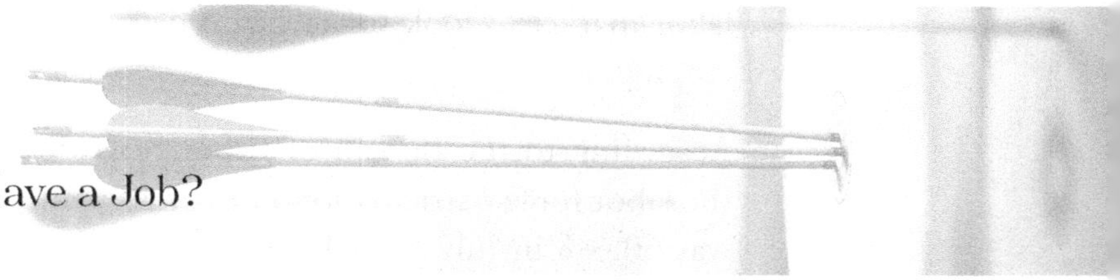

Investigation 15

How Many Can You Expect to Have a Job?

Sampling Distribution

Overview

This investigation explores the concept of a sampling distribution. Students will use simulation to model the sampling distribution of the sample number of successes by drawing slips of paper from a bag where 60% of the slips represent 16- to 24-year-olds with a job (success). Investigating the sampling distribution leads to the key idea that the mean of the sample number of successes will approximately equal the population expected number of successes.

This investigation follows the four components of statistical problem solving put forth in the *Guidelines for Assessment and Instruction in Statistics Education (GAISE) Report.* The four components are formulate a statistical question, design and implement a plan to collect data, analyze the data by measures and graphs, and interpret the results in the context of the original question. This is a GAISE Level C activity.

This investigation is based on lessons from *Probability Models* by Patrick Hopfensperger, Henry Kranendonk, and Richard Scheaffer, available as a free download at *www.amstat.org/ASA/Education/K-12-Educators.*

Instructional Plan

Brief Overview

» Read the scenario about the percent of teens who have a job.

» Discuss possible results of taking a random sample from a population with 60% of teens who have a job.

» Take a random sample of size 20 from a bag of slips of paper with 60% of the slips indicating teens have a job.

» Continue to take random samples to build a sampling distribution of the sample number of successes with sample sizes of 20.

» Describe the distribution and conclude the mean of the sampling distribution equals the expected number of successes (have a job).

Scenario

Many high-school and college students have a job after school or on weekends. Many work in a fast-food restaurant or as a clerk in a store.

Do you have a job? What kind of work do you do? Do you like your job?

If you don't have a job, do you think this is unusual?

The youth labor force of 16- to 24-year-olds working or actively looking for work increases sharply between April and July each year. During these months, large numbers of high-school and college students search for or take summer jobs, and many graduates enter the labor market to look for or begin permanent employment.

According to the Bureau of Labor Statistics, the labor force participation rate for all youth was 60.6% in July of 2017. (The labor force participation rate is the proportion of the civilian noninstitutional population that is working or looking and available for work. The civilian noninstitutional is made up of people 16 years and over residing in the US and not inmates or on active military duty.)
Source: www.bls.gov/news.release/youth.nr0.htm

Learning Goal

Investigate the sampling distribution of the sample number of successes through simulation.

Mathematical Practices Through a Statistical Lens

MP8. Look for and express regularity in repeated reasoning.

Statistically proficient students maintain oversight of the process, attend to the details, and continually evaluate the reasonableness of their results as they are carrying out the statistical problem-solving process.

Materials

Student worksheets are available at *www.statisticsteacher.org/statistics-teacher-publications/focus.*

» Cloth or paper bag

» Copies of the template run on stock paper; both sheets of the template for each group

» Student Worksheet 15.1 Data Collecting

» Student Worksheet 15.2 Template

» Exit Ticket

Estimated Time

Two 50-minute class periods. One period to collect the data and build a sampling distribution. A second period to analyze and draw conclusions pertaining to the sampling distribution.

Pre-Knowledge

» Students should be able to:

» Design and conduct a simulation

» Construct a dot plot

» Find the mean and standard deviation of a distribution using technology

Formulate a Statistical Question

Hand out Student Worksheet 15.1 Data Collecting. Have students work with a partner to answer questions 1 to 4.

Assume the labor force participation rate for all 16- to 24-year-olds in your city is 60%.

1. If your class is to select a random sample of 20 16- to 24-year-olds from the community, how many of the randomly selected 16- to 24-year-olds would you expect to be part of the labor force?

Answer: *Approximately 0.6(20), or 12 16- to 24-year-olds.*

2. If all the students in your class would each take a random sample of 20 16- to 24-year-olds from the city and ask each selected 16- to 24-year-old if they were working, do you think each class member would get the same number of 16- to 24-year-olds in the sample who are part of the labor force?

Answer: *It is unlikely they would all get exactly same result.*

3. Do you think it would be likely to find survey results of 13 out of the 20 16- to 24-year-olds, or 65%, reporting they have a job? Explain your answer.

Possible answer: *This result is likely; 13 is not much larger than the expected value of 12.*

4. Do you think it would be likely to find survey results of fewer than seven of the 20 16- to 24-year-olds, or 35% or less, reporting they have a job? Explain your answer.

Possible answer: *It is unlikely to get this result; seven is much smaller than the expected value of 12.*

Formulate a Statistical Question

Discuss the answers to questions 1 to 4. When discussing Question 2, encourage your students to begin to think about what they would expect for results. This leads into questions 3 and 4, in which the objective is to help students get an idea of what they think are likely and unlikely results. Consider having students give an interval of likely results.

Explain that the next step is to take many random samples of 20 slips of paper from a bag. Next, the students will investigate what results are likely and whether any patterns develop as they model the behavior of the results by building a sampling distribution of the sample number of successes (have a job).

Ask your students to consider the statistical question: "How many 16- to 24-year-olds out of 20 could have jobs if random samples of size 20 are taken from a population in which 60% of 16- to 24-year-olds have jobs?"

Collect Appropriate Data

Examine the template (Student Worksheet 15.2—two sheets each with 100 squares numbered 1 to 200) provided for this investigation. Note that the slips numbered from 1 to 120 are labeled with the word Job and those numbered 121 to 200 are labeled No Job. Cut out the slips and place the slips in a bag or container. Thoroughly mix the slips.

Tell your students the 200 slips of paper represent a population of 16- to 24-year-olds.

Explain that we are going to take a random sample of 20 slips of paper and count the number of slips that have the word Job written on them.

One way to collect this sample is to go around the room and have your students reach in the bag and select a slip. Note what the slip says. Continue selecting a slip until a sample of 20 slips has been selected. Count the number of slips that have the word Job written on them.

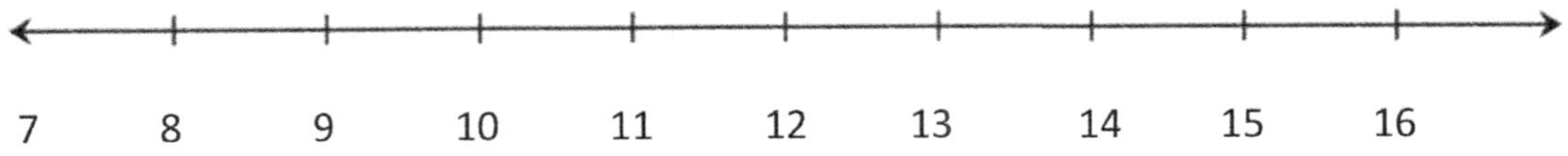

Figure 15.1: Number with a job from sample of 20

Ask your students what the results of the random sample of 20 slips represent.

Answer: *The results represent the number of 16- to 24-year-olds who said they had a job based on a random sample of 20 from a population of 16- to 24-year-olds in which 60% have a job.*

Ask your students if they will get the same number of slips with Job written on them if another random sample of 20 slips is selected?

Answer: *It is unlikely they would get the same result.*

Draw another random sample of 20 from the bag and find the number of 16- to 24-year-olds with a job. Again, ask your students what the results represent.

On the board or on poster paper, draw a number line like in Figure 15.1.

Give each group of students a bag and the template (Student Worksheet 15.2). Ask your students to cut out the slips of paper and place the slips in the bag. Have them thoroughly mix the slips.

Ask each group to take a random sample of 20 slips and record the number of successes (number that said Job) on the class dot plot.

Ask your students to repeat the simulation until the class has completed at least 50 trials.

Analyze the Data

Once students have completed their trials and reported their sample results on the class dot plot, point out to your students that the dot plot is an example of an approximate sampling distribution. Define a sampling distribution of the sample number of successes (students with a job) as a distribution of the sample number of successes from all possible samples of the same size from a population with a known proportion of successes (in this case, 60%).

Note: The total number of all possible samples of size 20 from 200 slips is ${}_{200}C_{20} = 1.61 \times 10^{27}$. Since it is not practical to create the sampling distribution of all possible samples, we will create an approximate sampling distribution.

Ask your students to answer questions 5 to 7. Discuss the student answers.

5. What should the title of our graph be? What did we graph?

Possible answer: *Simulated sampling distribution of the sample number of successes based on sample size = 20.*

6. Estimate the mean and standard deviation of the sampling distribution.

Possible answer: *Mean is approximately 11.5 and the standard deviation is approximately 1.8. (These are values for the example in Figure 15.2; the class data will vary from this example but be approximately the same.)*

7. Describe the shape of the sampling distribution.

Possible answers: *For this example in Figure 15.2, the shape is approximately symmetric and mound shaped.*

The sample results (Figure 15.2) are based on 68 student responses.

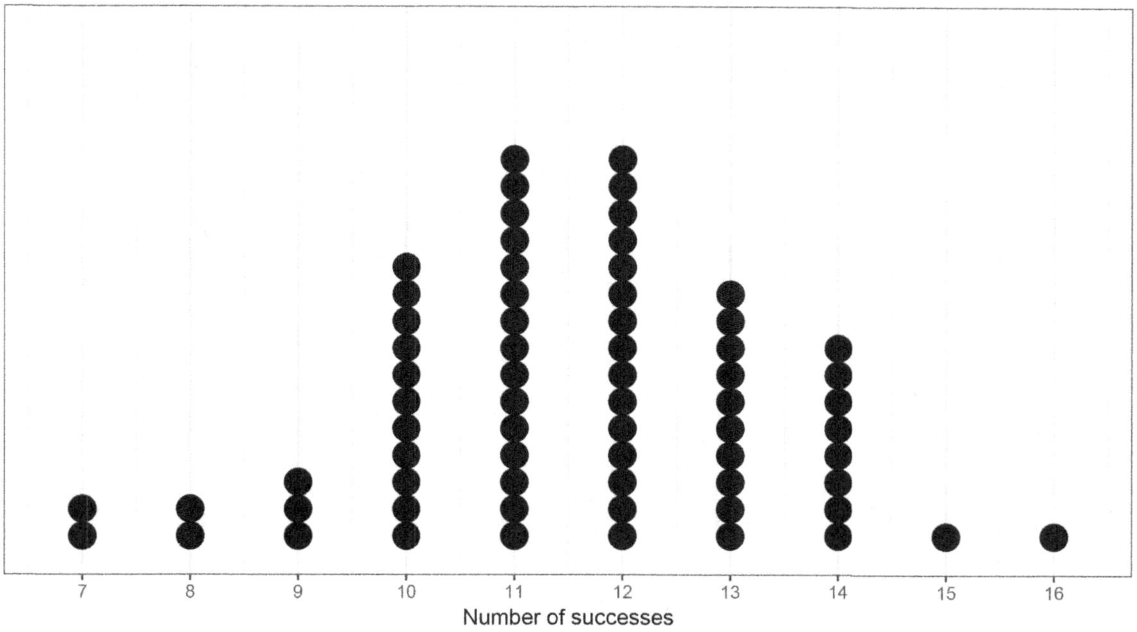

Figure 15.2: Dot plot of the results of 68 student responses

Ask your students to answer questions 8 to 11 on the worksheet.

8. Are you surprised the center of the distribution is close to 12?

Possible answer: *No, since the population proportion is 60%, this distribution should center around 0.60(20), or 12.*

9. If you took another random sample of 20 and found 10 said Job, would you call this a likely result? Explain.

Answer: *Ten is a likely outcome because 10 is close to the expected number of 12.*

10. If you took another random sample of 20 and found 15 said Job, would you call this a likely result? Explain.

Answer: *Fifteen is not a likely result. In our simulation, 15 or more was reported only twice out of 68 trials.*

11. If you took one more random sample of 20, give an interval you think would constitute a likely result? Explain.

Answer: *Between 9 and 14. This interval contained all the outcomes, except for six.*

Note: This answer is based on the dot plot shown in Figure 15.2.

Now explain that we want to change the *number* of successes to the *proportion* of students who have a job.

Add a second number line underneath the class dot plot, as shown in Figure 15.3.

Ask your students to change the number of successes to the proportion of students with a job from a sample of 20.

Answer: *Figure 15.3*

Ask your students to answer questions 12 and 13 on the worksheet.

12. Estimate the mean of the sample proportions.

Answer: *The mean will be approximately 0.57, or close to 0.60.*

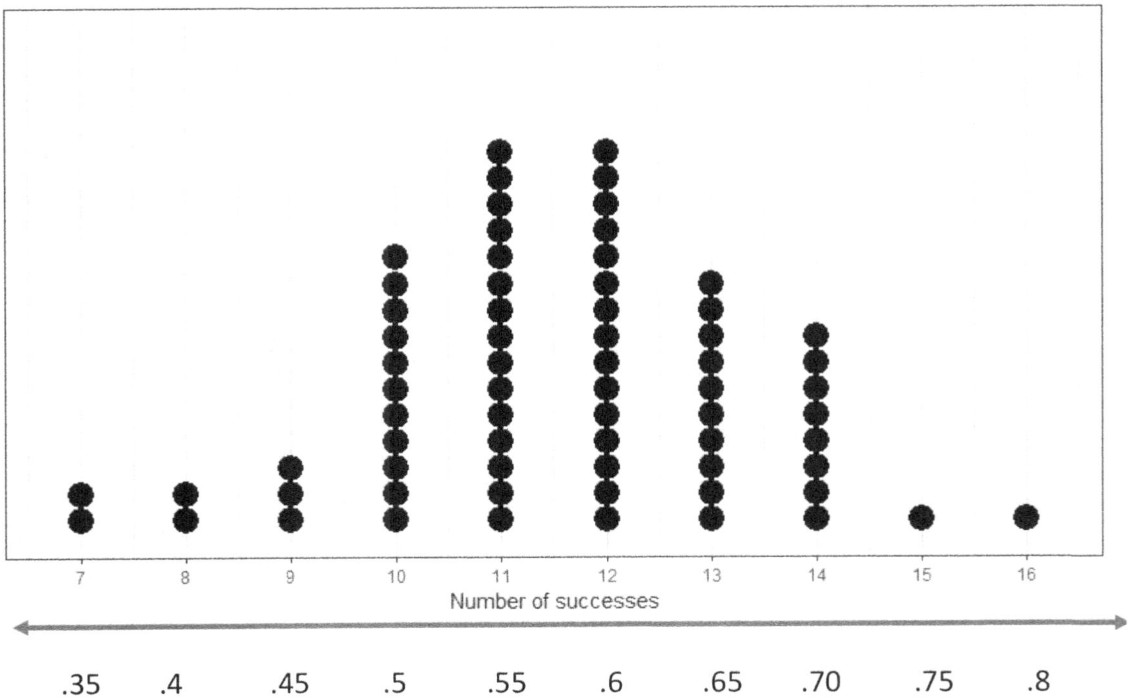

Figure 15.3: Dot plot of the proportion of students with a job from a sample of 20

13. What proportion would you expect for the mean? Explain.

Answer: The expected mean would be 0.60, the population proportion.

Interpret the Results in the Context of the Original Question

Based on the class simulation results, answer questions 14 to 16.

14. How many 16- to 24-year-olds out of 20 could have jobs if random samples of 20 are taken from a population in which 60% of 16- to 24-year-olds have jobs?

Answer: Between 9 and 14

15. What would be an unusual number to find through random sampling of 20 from a population in which 60% of 16- to 24-year-olds have a job?

Answer: Eight and fewer and 15 or more

16. Complete the following sentence:

The mean of the sample proportions will be equal to the value of the ________________.

Answer: population proportion

Additional Ideas

Search the Census Bureau website: (*www.census.gov*) for one of the following population proportions:

» Proportion of US households that subscribes to cable TV

» Proportion of US households that have a telephone landline as their only phone service

» Proportion of US residents over the age of 25 who are high-school graduates

Use this population proportion to develop a sampling distribution of the number of sample successes or sample proportions based on samples of size 20.

Exit Ticket

A random sample of 20 from 200 US households was taken and the number of cat owners was calculated.

This was repeated 75 times, and the dot plot of the results is shown in Figure 15.4.

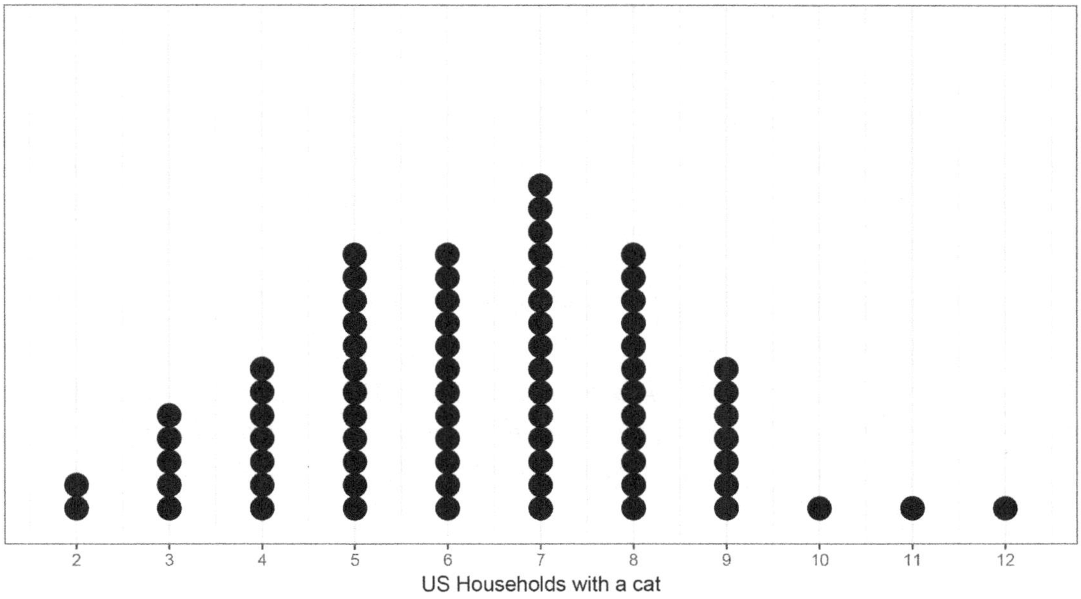

Figure 15.4: Dot plot of number of cat owners from a random sample of 20

1. Give a title to this graph.

Answer: *Simulated Sampling Distribution of the number of Cat Owners from a Sample Size of 20.*

2. Describe the shape and estimate the center and spread of the distribution.

Answer: *The distribution is mound shaped, centered at about 7, and has a spread (standard deviation) of about two.*

3. What is your estimate for the proportion of all US households that own a cat? Explain.

Answer: *The estimate is 7/20, or 0.35, the mean of the sampling distribution.*

Further Exploration and Extensions

1. Ask your students to find the standard deviation for the class results of the simulated distributions of the sample number of successes based on sample size of 20.

Answer: *In this example, the standard deviation is approximately 1.8.*

2. Have your students refer to the sampling distribution based on a sample size of 20. Find the percent of sample proportions within one standard deviation of the mean.

Answer: *In this example, with a mean of 11.5 and standard deviation of 1.8, the interval 9.7 to 13.3 would represent one standard deviation from the mean. Counting the dots between 10 and 13, inclusive, results in 51 out of 68, or about 75%.*

3. Have your students refer to the sampling distribution based on a sample size of 20. Find the percent of sample proportions within two standard deviations of the mean.

Answer: *In this example, with a mean of 11.5 and standard deviation of 1.8, the interval 7.9 to 15.1 would represent two standard deviations from the mean. Counting the dots between 8 and 1.5, inclusive, results in 65 out of 68, or about 95%.*

4. Could the Normal distribution be used to model the sampling distribution of the sample number of successes?

Answer: *The Normal distribution could be used to model the sampling distribution because the distribution is mound shaped and symmetrical. The proportion of data within one and two standard deviations is close to the theoretical results from the empirical rule.*

Investigation 16

Too Many Peanuts?

Investigating a Claim

Overview

This investigation introduces the concept of informal statistical inference. Using technology, students construct a sampling distribution of sample proportions to determine if an observed sample proportion would be considered unusual for a given population proportion. Students will be testing the claim that cans of mixed nuts contain approximately 50% peanuts. This investigation follows the four components of statistical problem solving put forth in the *Guidelines for Assessment and Instruction in Statistics Education (GAISE) Report*. The four components are formulate a statistical question, design and implement a plan to collect data, analyze the data by measures and graphs, and interpret the results in the context of the original question. This is a GAISE Level C activity.

Note: If any students have an allergy to peanuts, rather than open a can of nuts, use the data presented in the lesson.

Instructional Plan

Brief Overview

- Develop a statistical question about the proportion of peanuts in a can of mixed nuts that is claimed to contain approximately 50% peanuts.
- Prior to class, count the number of nuts in the can and consider that number to be the random sample size of mixed nuts taken from the population of all mixed nuts processed by the manufacturer. In this investigation, the number of mixed nuts is 258.
- Calculate the proportion of peanuts in the sample. In this investigation, the proportion of peanuts in the sample of 258 mixed nuts is 55% peanuts.
- Construct a sampling distribution of sample proportions of size (number of nuts in the can). In this investigation, 258 mixed nuts from a population with a proportion of 50% peanuts is used as an example.
- Based on the sampling distribution, find the probability of randomly obtaining a sample proportion of peanuts of at least 55% peanuts, assuming the population from which the sample is taken contains 50% peanuts.

Hand out Student Worksheet 16.1 Peanut Investigation.

Ask your students to read the scenario.

Scenario

Did you ever buy a can of mixed nuts and it seemed all you got in the can was peanuts and you were hoping for a lot of cashews and almonds?

A 1964 *Consumer Reports* investigation of 124 cans of mixed nuts, representing 31 brands bought in 17 American cities, determined that most mixed nuts at that time were mostly peanuts, often 75%. As of 1993, the Food and

Drug Administration (FDA) has required a container of mixed nuts to contain at least four varieties of tree nuts or peanuts. Each kind of nut must be present not less than 2% and not more than 80% of the number of nuts.

A major manufacturer of cans of mixed nuts makes the claim that their 10.3 oz. cans containing a mixture of peanuts, almonds, cashews, pecans, and Brazil nuts have approximately 50% peanuts.

As part of a statistics project, an 11th grader purchased a 10.3 oz. can of mixed nuts and found 142 peanuts in the can that contained 258 mixed nuts or approximately 55% peanuts.

Learning Goal

Use the sampling distribution of sample proportions and informally decide if a single sample proportion is unusual.

Mathematical Practices Through a Statistical Lens

MP3. Construct viable arguments and critique the reasoning of others.

Statistically proficient students use appropriate data and statistical methods to draw conclusions about a statistical question. They reason inductively about data, making inferences that take into account the context from which the data arose. They justify their conclusions and communicate them to others.

Materials

Student worksheets are available at *www.statisticsteacher.org/statistics-teacher-publications/focus.*

» 10.3 oz. can of mixed nuts that the manufacturer claims contains approximately 50% peanuts

» Statistical software or application to generate a sampling distribution of sample proportions. Possible applications: Graphing calculator with ProbSim app or computer software like GeoGebra or StatKey

» Student Worksheet 16.1 Peanut Investigation

» Exit Ticket

» Optional: Student Worksheet 16.2 StatKey Directions

Estimated Time

One 50-minute class

Pre-Knowledge

Students should be able to find the mean and standard deviation of a distribution using technology.

Does this mean the manufacturer's claim of approximately 50% peanuts is not correct? Does this provide convincing evidence that cans of mixed nuts from this manufacturer contain more than 50% peanuts?

Note: If appropriate, open the can of mixed nuts and count the total number of nuts and the number of peanuts. Determine the proportion of peanuts in the can. Use these results to complete this investigation.

Note: If there are students with peanut allergies, then use the 55% result computed from a can containing 258 nuts. This investigation will use the 55% results from a can containing 258 nuts as the example.

Formulate a Statistical Question

Start the discussion by explaining that we are going to assume the claim of approximately 50% of mixed nuts are peanuts is true. Another way to say this is the population proportion of peanuts in all the mixed nuts is approximately 50%. Explain that we are also going to assume the number of nuts in the can of mixed nuts represents a random sample from a population of all mixed nuts produced by this manufacturer.

Next, ask your students, "If the manufacturer's claim of 50% peanuts is true, how likely is it that we get a can of mixed nuts that contains 55% peanuts? Is this an unusual result? What is the probability we could get a sample containing 55% peanuts by chance from a population containing 50% peanuts?"

Ask your students to consider the statistical question: "Assuming the manufacturer's claim that a can of mixed nuts contains 50% peanuts is true or the population proportion is 0.5, is the proportion of peanuts of 0.55 found in a can (sample) of mixed nuts an unusual result?"

Collect Appropriate Data

Ask your students to answer questions 1 to 3.

We are going to assume the population proportion of peanuts in all the mixed nuts processed by the manufacturer is 50% and the sample of 258 nuts (one can) was a random sample of all the mixed nuts produced by the manufacturer.

1. Assuming the claim that 50% of a can is peanuts, how many peanuts would you expect to be in a can of 258 nuts?

Answer: Fifty percent of 258 is 129.

2. If a random sample of 258 mixed nuts yielded 134 peanuts, would you think it an unusual result? Why or why not?

Answer: This would not be considered unusual, since 134/258 is approximately 52%, which is close to 50%.

3. If a random sample of 258 mixed nuts yielded 155 peanuts, would you think it an unusual result? Why or why not?

Answer: This is a very unusual result. 155/258 is about 60%, which is much higher than the claim of 50%.

Ask your students to complete questions 4 and 5.

Note: The following results were constructed using the statistical computer application-StatKey (*www.lock5stat.com/StatKey*).

Steps for using StatKey are on Student Worksheet 16.2 StatKey Directions.

4. As directed by your teacher, use statistical software to construct a simulated sampling distribution of at least 200 sample proportions based on a sample size of 258—number of nuts in the can—and assuming a population proportion of 50%.

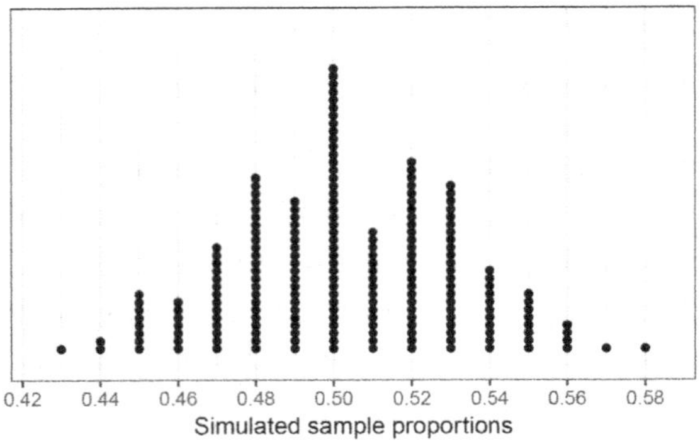

Figure 16.1: Sampling distribution of 200 sample proportions

Sample answer: *Figure 16.1*

Analyze the Data

Ask your students to answer questions 5 to 9.

5. What do you expect the mean of the simulated sampling distribution of sample proportions to equal?

Answer: Approximately 0.50

6. Using statistical software, find the mean and standard deviation of the simulated sampling distribution.

Sample answer: Mean = 0.501, or 0.5; standard deviation = 0.030

Note: The standard deviation of the sampling distribution is called the standard error of the sample proportion.

7. Describe the simulated sampling distribution of the sample proportions.

Sample answer: Mound shaped or approximately Normal with a mean of approximately 0.5 and a standard deviation of approximately 0.03. Most sample proportions are between 0.44 and 0.56.

8. Count the sample proportions on the plot that are greater than or equal to the proportion of peanuts in the class can (0.55 in this example). How many sample proportions were greater than or equal to the class proportion of peanuts?

Sample answer: 14 out of 200 (based on this example)

9. Estimate the probability of the class getting a can of mixed nuts and obtaining a sample proportion of __% (in this example, 55%) peanuts or greater from a population with the population proportion equal to 0.50 peanuts.

Sample answer: 14/200 or 7% chance (based on this example)

Interpret the Results in the Context of the Original Question

Ask the students to answer questions 10 to 12.

10. Do you think the proportion of peanuts in the class can of mixed nuts was an unusual result assuming the manufacturer's claim of 50% is correct?

Answer: Since an estimate for the probability of obtaining the class sample proportion is 7% (answer and interpretation will vary based on the class results), the sample proportion is not that unusual. There is "some" evidence that the number of peanuts is higher than the usual amount, but there is not enough evidence to say the company's claim of 50% peanuts is not true.

11. What proportion of peanuts would you consider to be an unusual result? Based on the simulated sampling distribution, what is an estimate for the probability of obtaining that proportion or more by chance?

Possible answers: Answers will vary, but students may respond with a sample proportion of around 57% or higher. In this example, the probability is approximately 2/200, or 0.01.

12. If you got such a can (high proportion of peanuts), would you have reason to believe the manufacturer's claim is not correct?

Possible answer: *Even though it can happen, the probability is very low, and I would not believe the manufacturer claim of 50% peanuts.*

Additional Ideas

» Use survey results from your class or classes and test the claim that 75% of teens use Snapchat.

» Use survey results from your class or classes and test the claim that 50% of teens use Twitter.

» Use survey results from your class or classes and test the claim that fewer than 30% of teens use Tumblr, Twitch, or Linkedln.

Exit Ticket

The American Society for the Prevention of Cruelty to Animals (ASPCA) claims that approximately 35% of US households have at least one cat. Assuming the ASPCA's claim is correct, a sampling distribution of 100 sample proportions based on a sample size of 50 and population proportion of 0.35 is shown in Figure 16.2.

Mean of simulated sampling distribution = 0.35 and a standard deviation = 0.06

1. Describe the simulated sampling distribution of the sample proportion.

Answer: *The distribution is mound shaped or approximately Normal with a mean of 0.35 and standard deviation of 0.06.*

2. A random sample of 50 US households found the proportion of households with at least one cat was 0.22. Mark the sample result of 0.22 on the dot plot. How many sample proportions are less than or equal to the sample result of 0.22?

Answer: *Approximately 3 out of 100, as shown in Figure 16.3*

3. What is an estimate for the probability of obtaining a sample proportion of 0.22 or less from a population with 0.35 households with a cat?

Exit Ticket Cont.

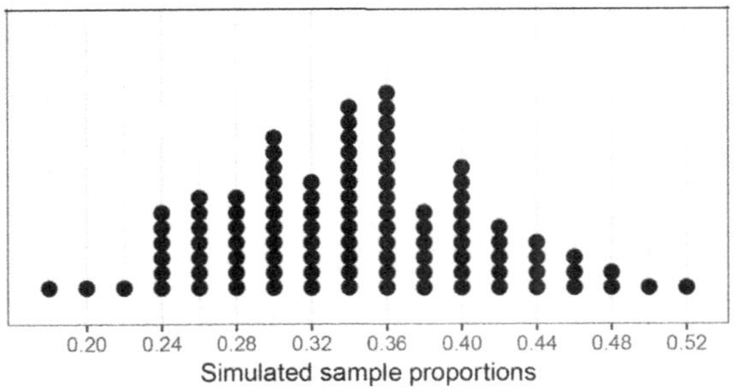

Figure 16.2: A sampling distribution of 100 sample proportions based on a sample size of 50 and population proportion of 0.35

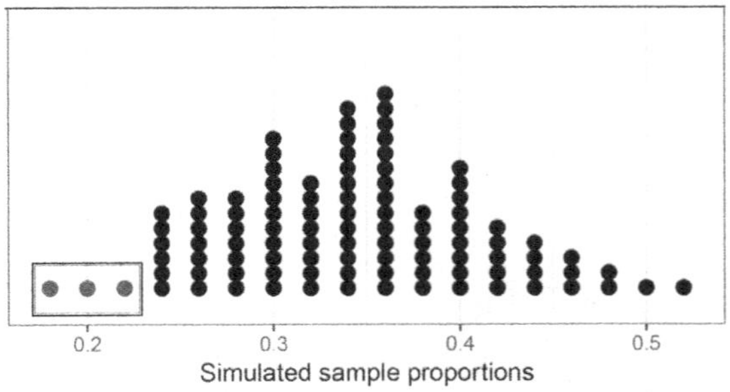

Figure 16.3: Sample result

Answer: *Approximately 3/100 or 0.03*

4. Do you think the proportion of households with a cat (22%) was an unusual result, assuming the ASPCA's claim of 35% is correct? Explain your answer.

Possible answer: *The probability of obtaining a sample result of 0.22 households with a cat from a population with a proportion of 0.35 households with a cat is about 3%. This is a low probability, so I think this is an unusual result.*

Further Exploration and Extensions

1. Introduce the *p*-value.

   The 7% (peanut example) is called a *p*-value. Assuming the company's claim of 50% peanuts in the can is correct, the *p*-value is the probability of getting the results you did (or more extreme) purely by chance.

   A *p*-value less than or equal to 5% is considered statistically significant, to where the researcher would reject the assumption that the observed results were due to random variation and conclude there is strong evidence to support that the results indicate the claim is not true.

   The concept of a *p*-value was formally introduced by Karl Pearson, in his Pearson's Chi-Squared test. The use of the *p*-value in statistics was popularized by Sir Ronald Fisher. In his book *Statistical Methods for Research Workers* (1925), Fisher proposed the level $p = 0.05$ as a possible limit for statistical significance and applied this to a Normal distribution, thus yielding the rule of two standard deviations (on a Normal distribution) for statistical significance using the empirical rule, or 68–95–99.7 rule.

2. Activity to illustrate the general rule that a *p*-value of less than or equal to 5% is considered statistically significant. *Note:* This activity usually takes a few minutes to complete.

   » You will need a deck of all red cards. All the cards need to have the same design on the front so they look like a regular deck of cards. Have the cards in the box so the students assume the box contains the normal arrangement of 26 red and 26 black cards.

   » Tell the students you are going to randomly divide them into two groups based on the color of the card. Red card they are in Group 1 and black card in Group 2.

   » Remove the cards from the box and carefully shuffle them without the students seeing any of the red color on the back of the cards.

   » Go to the first student and turn over a card. Since it will be red, tell the student s/he is in Group 1.

   » Go to the second student and turn over a card. That student will also be in Group 1.

   » Continue until the students get suspicious, usually around the fourth or fifth student.

Further Exploration and Extensions Cont.

» Once they get suspicious, stop and discuss the results:

» Why did you get suspicious?

Answer: They might say too many reds in a row.

» What did they expect to happen? Why did they expect this?

Answer: Expect about half the cards to be red and the others black. If this were a regular deck of cards, we would expect half to be red and half to be black.

» What is the probability that we would get this many red cards in a row, assuming this was a regular deck of cards?

Answer: 4th student – $(1/2)^4$*, about 0.06; 5th student –* $(1/2)^5$*, about 0.03*

Point out that they got suspicious when the probability of observing five red cards in a row was about 3%. They assumed the deck was a regular deck of cards—that is, the probability of turning over a red card was 50% and they reacted (the results they saw were unusual) between 6% and 3%.

Note: The process followed is comparable to what is done in traditional hypothesis testing. Hypothesis testing refers to the formal procedures used by statisticians to reject or fail to reject the null hypothesis. In the flipping card example, the null hypothesis is the probability of turning over a red card is 50%. We assumed the null hypothesis (probability of red 50%) is true. Given the results of the card-turning simulation, we decided to reject the null hypothesis and conclude the probability of turning over a red card is not 50%.

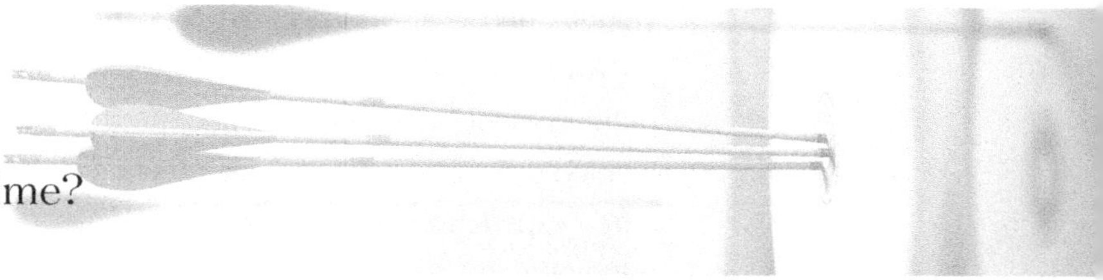

Investigation 17

How Many Hours of Volunteer Time?

Bootstrapping

Overview

This investigation develops an interval estimate for a population mean through a resampling method called bootstrapping. Bootstrapping is a technique in which a large number of random samples of the same size are repeatedly drawn, with replacement, from a single original sample. The interval that includes the middle 95% of the resampled sample means forms a bootstrap interval.

This investigation follows the four components of statistical problem solving put forth in the *Guidelines for Assessment and Instruction in Statistics Education (GAISE) Report*. The four components are formulate a statistical question, design and implement a plan to collect data, analyze the data by measures and graphs, and interpret the results in the context of the original question. This is a GAISE Level C activity.

Note: Before having your students read the scenario, discuss with them that some high schools require students to perform a minimum number of community service hours to graduate. Some high schools also allow students to earn credit toward their high-school graduation through community service.

Hand out Student Worksheet 17.1 Volunteer Work. Have your students read the scenario.

Scenario

Does your high school have a requirement of students to perform community service hours? If there is a requirement, how many hours are required to fulfill the responsibility? Do you already volunteer your time? What type of volunteer work do you do? How many hours do you volunteer? What are the benefits of volunteering?

There are many volunteer opportunities available for high-school students to take part in.

Some places one might volunteer include a hospital, nursing home, animal shelter, food bank, library, tutoring center, museum, beach or park, or church. The excerpt below discuses benefits for high-school students who volunteer.

> Volunteering has many benefits. Through volunteering, you'll get to explore a passion you have (such as literature or medicine). Also, by volunteering, you can support a cause you love such as helping the homeless. You can also meet like-minded students, who share your passion or want to support that cause.

> Volunteering is a great opportunity to test out whether you'd like to pursue a specific career (such as medicine, education, etc.). It's great to try and find your passion in high school, so you don't waste time and money during college trying to figure out what you want to major in. If you don't enjoy volunteering at a hospital, maybe pre-med isn't for you. If you love volunteering at an animal shelter, maybe you should pursue a career as a veterinarian.

Volunteering is also a great extracurricular for your college application. It shows you selflessly dedicated your time and effort to helping others! Additionally, volunteering is a free experience that won't cost you anything other than time.

Source: https://blog.prepscholar.com/volunteer-opportunities-for-teens

Formulate a Statistical Question

A local school board is considering adding a community service graduation requirement for all district high-school students. To help the school board make an informed decision, a small group of statistics students decided to select a random sample of district high-school students who are already volunteering to determine the type of volunteer service and how many

Learning Goal

Understand the concept of bootstrapping and use bootstrapping to construct an interval estimate for a population mean.

Mathematical Practices Through a Statistical Lens

MP5. Use Appropriate Tools Strategically

Statistically proficient students can use technological tools to carry out simulations for exploring and deepening their understanding of statistical and probabilistic concepts.

Materials

Student worksheets are available at *www.statisticsteacher.org/statistics-teacher-publications/focus.*

» Student Worksheet 17.1 Random Sample of 50 Hours (copies run on stock paper)

» Student Worksheet 17.2 Volunteer Work

» Cloth or paper bag

» Statistical software or application that can generate a sampling distribution of sample means using a bootstrap method (possible application: StatKey, *www.lock5stat.com/StatKey*)

» Optional: Student Worksheet 17.3 StatKey Directions

» Exit Ticket

Estimated Time

One to two 50-minute class periods. One period to read the scenario and collect data. A second period to discuss bootstrapping and analyze the collected data and interpret the results.

Pre-Knowledge

Students should be able to use technology to construct a dot plot and find the mean and standard deviation of a sampling distribution.

Table 17.1

10	13	13	7	10	46	23	21	30	41	18	23
17	27	44	31	83	81	59	111	12	12	23	118
182	124	101	262	349	89	68	50	63	350	249	271
311	10	27	45	19	36	311	486	503	33	42	20
29	31										

hours the students are volunteering. They decided to investigate the statistical question: "For district high-school students who volunteer, what is an interval estimate for the mean number of hours they volunteer per year?"

Collect Appropriate Data

Explain that the 50 values on Student Worksheet 17.2 Random Sample of 50 hours represent the number of hours per year that 50 randomly selected district high-school students reported they volunteer (see Table 17.1). The 50 students were randomly selected from a large group of district students who reported they volunteered during the past year.

Explain that the school board would like to use these sample data to make an inference about the number of hours all district high-school students volunteer.

Ask your students for any observations or questions they have about the data.

Possible answer: *There is a large spread in the data going from 7 to more than 500.*

Ask your students to answer questions 1 to 4.

1. Construct a dot plot of the 50 times.

Answer: *Figure 17.1*

2. Find the mean of the distribution of times and describe the distribution.

Answer: *98.68 hours. The distribution is skewed to the right, with much of the data clustered between 0 and 50 hours. There are six times that are greater than 300 hours. Note: Standard deviation is 126.7 hours.*

3. What would happen if another random sample of 50 district students were taken?

Answer: *We would get different results but likely a similar looking distribution.*

4. What patterns would emerge if a large number of random samples of 50 students were taken and sample means were used to build a sampling distribution?

Answer: *The distribution of sample means would be approximately mound shape with a mean approximately equal to the population mean.*

Explain that the issue is that the school board only has the one sample of 50 times listed above. It is time consuming and usual-

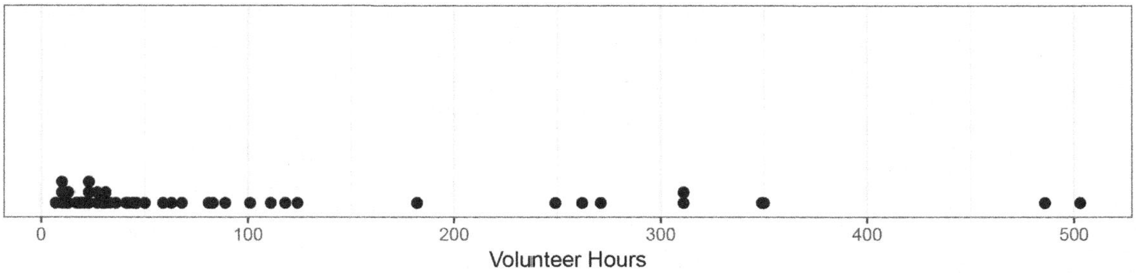

Figure 17.1: Dot plot of the 50 times high-school students reported they volunteer

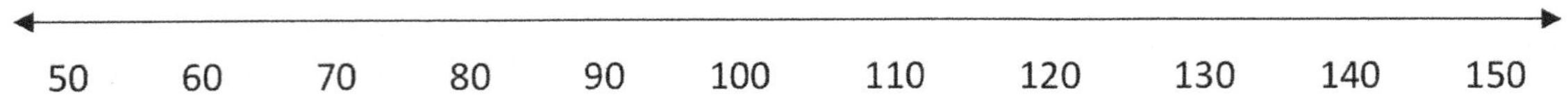

Figure 17.2: Number line of sample means

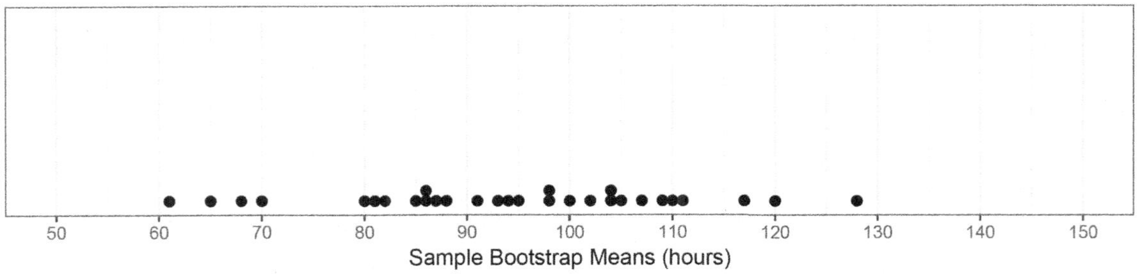

Figure 17.3: Number line of sample means

ly difficult to take a large number of random samples. In this investigation, we are going to use the single random sample result as if it is a population where the mean is approximately 99 hours. To help the school board draw conclusions about the number of volunteer hours, we are going to use a technique called bootstrapping. This technique uses the random sample of 50 in place of the actual population distribution of volunteer hours. That is, we will think of the distribution we have as being an estimate of the population distribution of volunteer hours. The bootstrapping method takes repeated samples of the same sample size with replacement and creates a sampling distribution of the sample bootstrap means.

Note: The use of the term bootstrap derives from the phrase to pull oneself up by one's bootstraps. You're trying to pull yourself up from what you've got. In a data sense, you're going to use the sample data itself to try to get more information about a population mean—the mean number of hours district high-school students volunteer.

Hand out Student Worksheet 17.1 Random Sample of 50 Hours.

Cut out the slips of paper with the 50 times and place the slips in a bag or container. Thoroughly mix the slips.

Explain that we are going to take a random sample (with replacement) of 50 slips. Go around the room and have a student draw out a slip, record the outcome, and return the slip to the bag. Continue until you have a total of 50 observations. This is called a bootstrap sample.

Note: It is possible you will draw each of the 50 slips exactly once, but this is highly unlikely. It is more likely you will draw some slips more than once and some slips not at all.

Find the mean of this bootstrap sample. This is a bootstrap estimate of the population mean.

Now have your students cut out the slips with the 50 numbers and place the slips in a bag or container. Have them thoroughly mix the slips.

5. Ask students to take a random sample of 50 slips of times with replacement and find the mean of their sample.

Place a number line on the board or poster paper similar to the one shown in Figure 17.2.

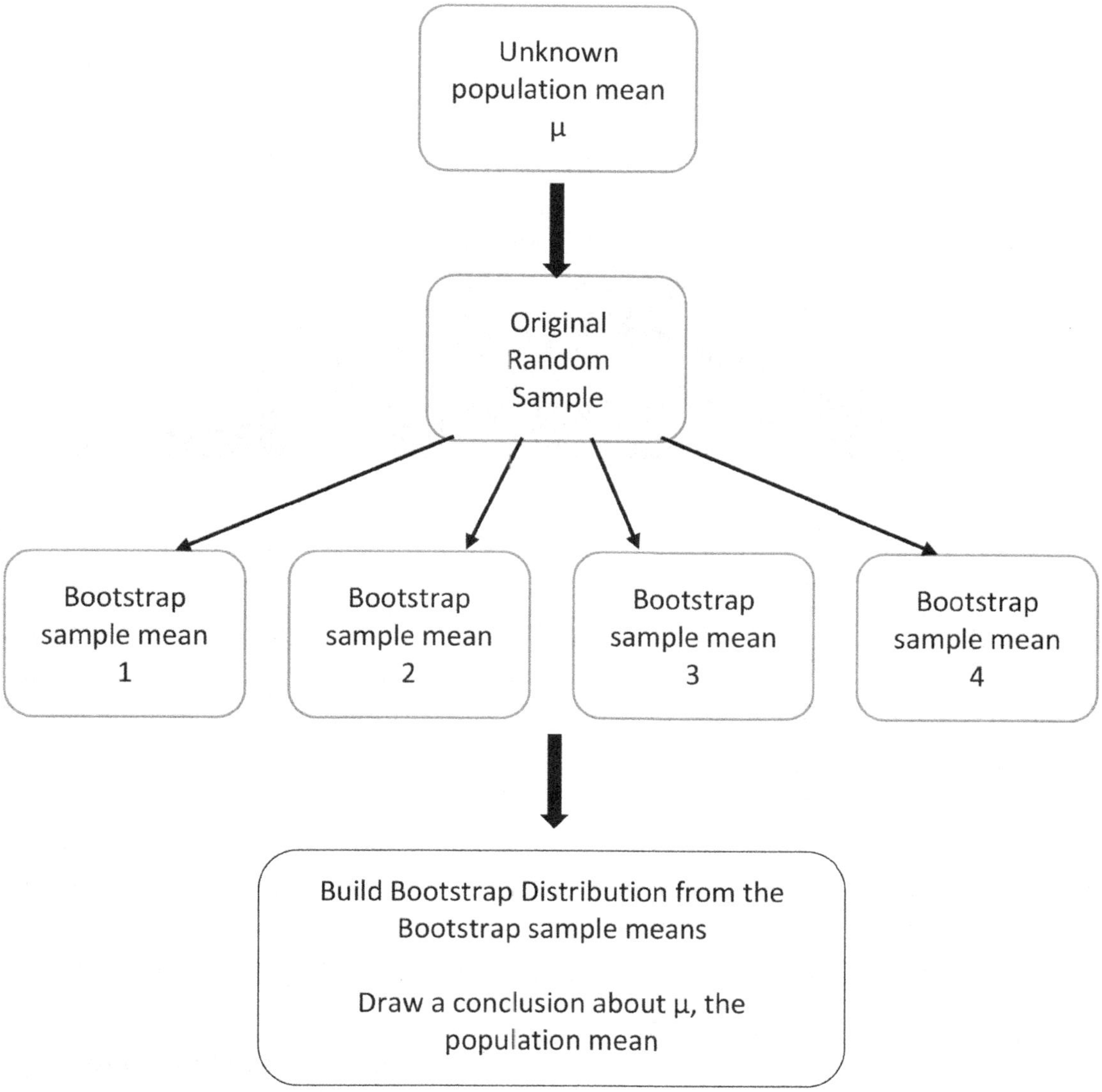

Figure 17.4: Bootstrapping method

Have your students record their sample means on the number line.

Answer: Sample dot plot (Figure 17.3) of 30 bootstrap sample means.

Note: Here is a diagram (Figure 17.4) that could be used to help explain the bootstrapping method. The symbol μ represents the population mean or the actual mean number of hours for all the student volunteers.

Analyze the Data

6. Using the class distribution of sample bootstrap means, find the mean of the distribution.

Answer: The mean will be approximately 98 or 99 hours.

Explain that it would be helpful to have more bootstrap sample means so we could get a clearer picture of the sampling distribution.

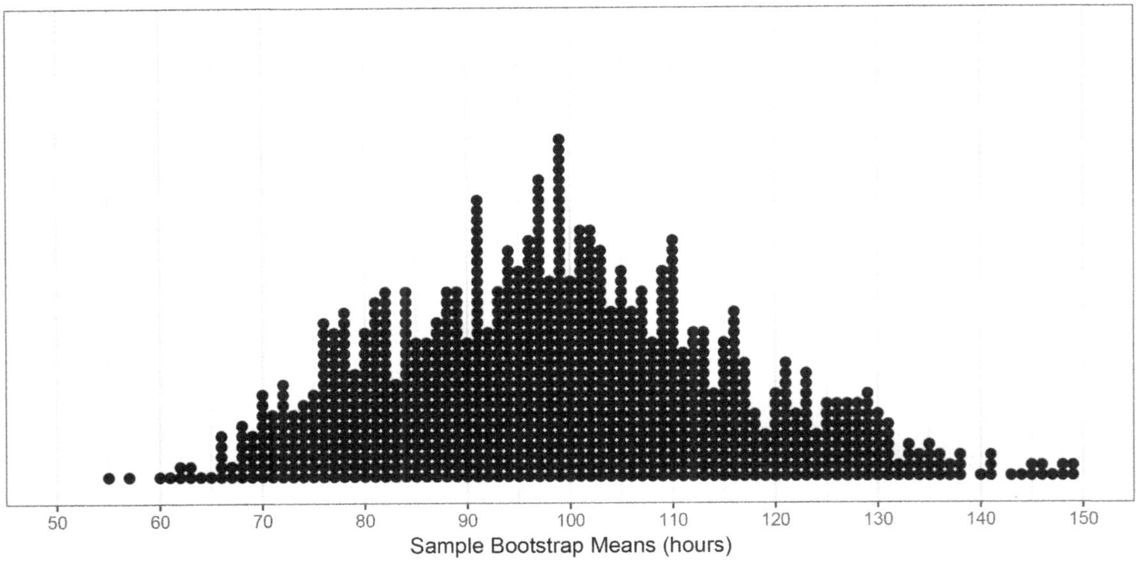

Figure 17.5: 1000 bootstrap sample means

So, rather than continuing to take random samples of 50 from the bag, use statistical software to generate a large number of bootstrap samples and display the sampling distribution of the sample means from the bootstrap samples.

Note: Worksheet 17.3 StatKey Directions contains the directions for using StatKey (*www.lock5stat.com/StatKey*) to generate the bootstrap samples. If students have access to computers, it is recommended they use software and construct a bootstrap confidence interval.

7. Use the statistical software and generate 1000 bootstrap sample means.

Possible answer: *Figure 17.5*

Rather than give just the mean, it would be helpful for the school board to have an interval where the population mean (volunteer hours of all the district students) would likely fall.

8. Using the sample distribution of the 1000 bootstrap sample means, between which two sample means is approximately the middle 95% of the distribution?

Possible answer: *Since there are 1000 sample means, we could eliminate the bottom 2.5%, or the lowest 25 sample means, and eliminate the top 2.5%, or the highest 25 sample means. See Figure 17.6.*

The two sample means 69 and 135 form an interval between which 95% of the bootstrap sample means are located. This interval gives an interval where the population mean is likely to be.

Interpret the Results in the Context of the Original Question

Ask the students to complete questions 9 to 12.

9. Using the results from the bootstrapping resample method, answer the original statistical question, "For district high-school students who volunteer, what is an interval estimate for the mean number of hours they volunteer per year?"

Possible answer: *The true mean number of hours of all the district high-school students who perform volunteer work is approximately between 68 and 136 hours, exclusive.*

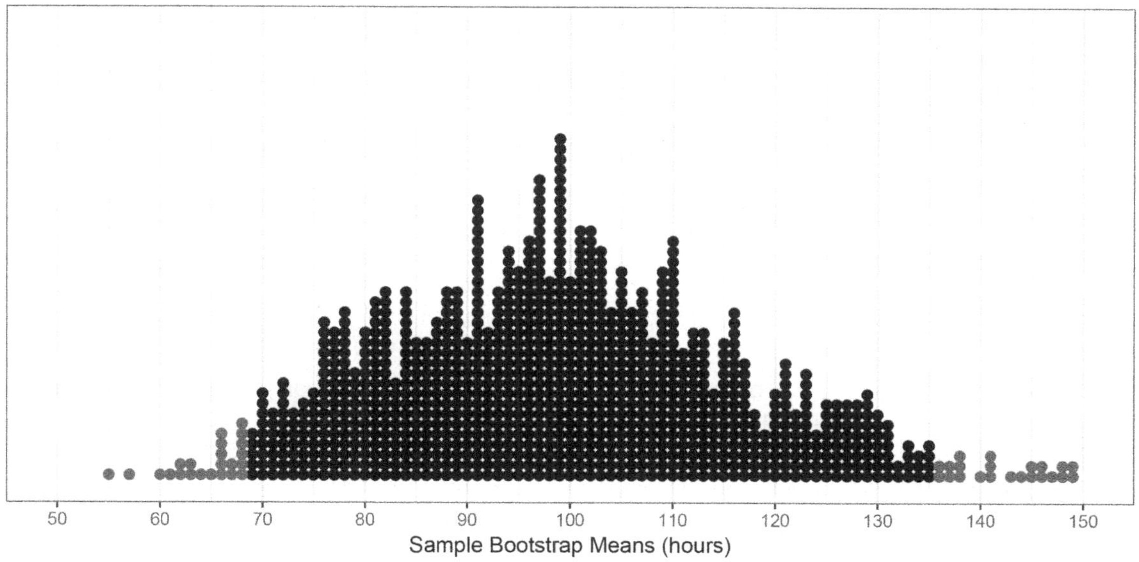

Figure 17.6: 1000 bootstrap sample means with lowest and highest sample means shaded lighter

10. Share your interval with others in class. Compare the intervals and discuss the similarities and differences.

Possible answer: *Most of the intervals will be close to the same.*

11. Write a brief summary of the bootstrap method and how it works.

Possible answer: *First, a random sample is taken from a large population. This random sample is used as representing the whole population. Many random samples are taken with replacement from the original sample, and the distribution of the sample means is created. This distribution is used to construct an interval estimate of the middle 95% of the sample means, which represents the actual value of the population mean.*

12. Explain how the bootstrap method could be used to construct an interval estimate for the middle 90% of the distribution.

Possible answer: *Take the bottom 5% off and the top 5% off the distribution of bootstrap sample means.*

As a final discussion, emphasize to your students that the interval we created is really dependent on how well the original random sample of 50 represents the population. If the original sample was biased or too small to truly represent the population distribution of volunteer hours, then the bootstrap method won't produce a valid result.

Additional Ideas

In Investigation 2: Are Baseball Games Taking Longer?, random samples of the length of Major League (MLB) Baseball games in three years (1957, 1987, and 2017) were given. Use each of the years and generate a sampling distribution of bootstrap sample means and construct a 95% interval estimate for the actual population mean (true mean length of all the games played during that year). Compare the three intervals and investigate whether the average length of MLB games is getting longer.

Exit Ticket

Chicago has been keeping records for the number of inches of snow for many years. Forty years were randomly selected from all the years snowfall has been recorded, and the amount of snowfall for those years formed a random sample.

This random sample of 40 snowfall amounts was used to take 100 bootstrap samples, and the sample bootstrap means are shown in the dot plot in Figure 17.7.

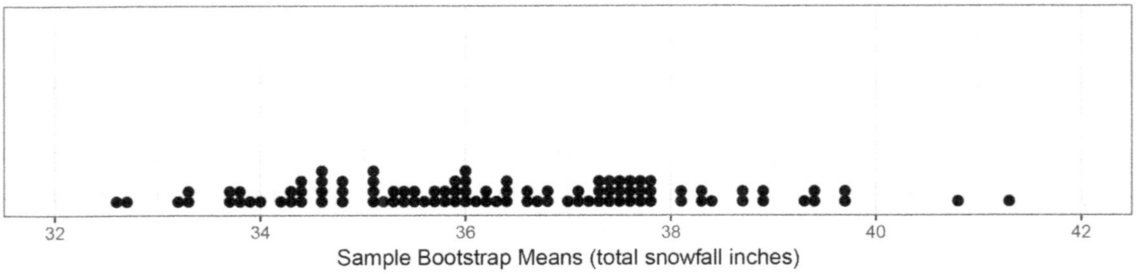

Figure 17.7: Sample bootstrap means of 40 snowfall amounts

1. What is an estimate for the mean number of inches of snow in Chicago over all the years snowfall has been recorded?

***Answer:** Approximately 36 inches—the mean of the sampling distribution.*

2. What is a 95% bootstrap interval estimate for the mean number of inches of snow in Chicago for all the years snowfall has been recorded? Explain how you got your answer.

***Answer:** Approximately 33 to 41 inches. Since there are 100 sample means, eliminate the lowest 2.5%, or the lowest two or three, and eliminate the highest 2.5%, or the top two or three.*

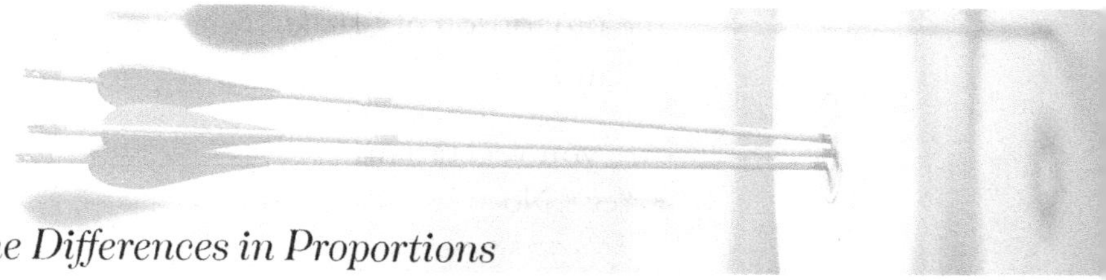

Investigation 18

How Stressed Are You?

Exploratory Lesson: Comparing the Differences in Proportions

Overview

This investigation offers two options for students. One option provides students an opportunity to use the four components of statistical problem solving by designing their own investigation around a topic of interest that involves exploring whether two proportions are significantly different. Several suggestions are included in this investigation, and students could be encouraged to come up with their own questions of interest. Encourage students to work in pairs or small groups.

The results could include written and oral presentations and/or construction of a poster to display the data and answer the statistical question. Information about creating a statistical poster, a rubric, and competition information can be found at *www.amstat.org/asa/education/ASAStatistics-Poster-Competition-for-Grades-K-12.aspx.*

A second option provides the set of directions for an investigation titled, "American Teens Compared to New Zealand Teens." This option is suggested for students who may need more scaffolding and direction when designing a simulation and analyzing the simulation results. This option is based on Lesson 14 from *Making Sense of Statistical Studies,* published by the American Statistical Association and available at *ww2.amstat.org/education/msss/index.cfm.*

This investigation follows the four components of statistical problem solving put forth in the *Guidelines for Assessment and Instruction in Statistics Education (GAISE) Report.* The four components are formulate a question, design and implement a plan to collect data, analyze the data by measures and graphs, and interpret the results in the context of the original question. This is a GAISE Level B or C activity, depending on the amount of scaffolding provided.

Instructional Plan

Instructions for Designing Your Own Investigation

Explain that students will follow the four steps of the statistical problem-solving process. Have students work in pairs or small groups. Distribute Student Worksheet 18.1 Directions for Student-Designed Investigation.

Formulate a Statistical Question

Students will brainstorm topics that might interest their group that include a categorical variable. Your students could design a question and take a random sample of two groups in their high school, such as freshmen and seniors or students and teachers. Students could model their question based on the questionnaire on the Census at School website or design questions based on a topic of student interest.

Some possible questions to explore based on ideas from the Census at School website (*ww2.amstat.org/censusatschool/students.cfm*) include the following:

» Is the proportion of students who are vegetarians significantly different from the proportion of teachers who are vegetarians?

22. Are you vegetarian?
 □Yes □No

» Is the proportion of students who drink energy or sports drinks significantly different from the proportion of teachers who drink energy or sports drinks?

24. What type of beverage do you drink most often during the day?
□Water □Soft drink (caffeinated) □Tea □Milk □Soft drink (non-caffeinated) □Coffee □Juice □Energy drink □Sports drink □Powdered drink (e.g., Kool-Aid, Tang)

» Is the proportion of teachers who like country music significantly different

Learning Goal

Understand what it means for two proportions to be significantly different.

Mathematical Practices Through a Statistical Lens

MP1. Make sense of problems and persevere in solving them.

Statistically proficient students understand how to carry out the four steps of the statistical problem-solving process: formulating a statistical question, designing a plan for collecting data and carrying out that plan, analyzing the data, and interpreting the results.

Materials

Student worksheets are available at *www.statisticsteacher.org/statistics-teacher-publications/focus.*

- Student Worksheet 18.1 Directions for Student-Designed Investigation
- Student Worksheet 18.2 Difference Between Proportions
- Student Worksheet 18.3 Template

Estimated Time

One to three 50-minute class periods, depending on final report and amount of work required outside of class.

Pre-Knowledge

Students should be able to:

- Summarize data using a dot plot
- Find the mean and standard deviation
- Conduct a simulation to construct a sampling distribution
- Use a simulated sampling distribution to determine what values are unlikely outcomes

than the proportion of students who like country music?

36. What is your favorite type of music? Select one.
□Classical □Pop □Rhythm and blues (R&B) □Other □Country □Punk rock □Rock and roll □Heavy metal □Rap/Hip hop □Techno/Electronic □Jazz □Reggae □Gospel

» Is the proportion of females who would like to be able to fly significantly different than the proportion of males who would like to fly?

37. Which of the following superpowers would you most like to have? Select one.
□Invisibility □Telepathy (read minds) □Freeze time □Super strength □Fly

Is the proportion of 9th graders who would give $1000 to environmental causes significantly different than the proportion of 12th graders who would give $1000 to environmental causes?

40. If you had $1000 to donate to a charity of your choice, what type of organization would you choose?
□Arts, culture, sports (e.g., community centers, museums, sports teams, music programs)

□Health (e.g., cancer, AIDS, diabetes research)

□Religious (e.g., church or activities related to worship)

□Environmental (e.g., saving forests, clean air, clean water)

□Wildlife, animals (e.g., endangered species, prevention of cruelty to animals)

□Education/Youth development (e.g., reading, literacy and skills training, after-school programs)

□International aid (e.g., disaster relief, health, education and food aid in poor countries)

Students should then develop a statistical question. Students are expected to check in for approval at this point before moving on to collecting data.

Collect Appropriate Data

Students can either take a random sample from the database at the Census at School website or take a random sample of the appropriate groups in their school. Direct students to outline the data-collection process, including possible complications and how these might be handled.

Once the data are collected, direct students to design a two-way table and find the difference between the proportions of interest. They will then design a simulation similar to the simulation outlined in Investigation 11 or the second part of this investigation based on the assumption that there is no difference between the two proportions. Students should run a large number of trials. For each trial, they should record the simulated difference between the two proportions.

Analyze the Data

Data analysis should include a dot plot and summary of the simulation.

Interpret the Results in the Context of the Original Question

Interpret the analysis of the data in the context of the situation. Be sure to answer the statistical question and support the answer with the data analysis.

Option 1: Write and orally present a report summarizing your results. Your report and presentation should include the following:

» The statistical question investigated and why it was chosen

» A description of the population sampled

» A summary of the data collection

» The collected data, organized as appropriate

» Analysis and descriptions of the data, using calculations, tables, graphs, and plots. Note any unusual results.

» Conclusions about the statistical question

» Recommendations for any follow-up studies or questions that may be investigated

Option 2: Create a data visualization poster and orally present the poster summarizing your results.

» The poster should include the following:

» The statistical question as the title of the poster

» The organized collected data—tables and graphs

» Conclusions about the statistical question

The oral report should include the following:

» The reason the statistical question was chosen

» A description of the populations sampled

» A summary of the data collection

» Analysis and descriptions of the data, using calculations, tables, graphs, and plots. Note any unusual results.

» Recommendations for any follow-up studies or questions that may be investigated

Instructions for "American Teens Compared to New Zealand Teens"

Scenario

The World Happiness Report is a survey of the state of global happiness. The World Happiness Report 2018 ranks 156 countries by their happiness levels. The report named Finland as the top-ranked—happiest—country in the world. New Zealand ranked eighth, and the United States ranked 18th. All the top countries tend to have high values for all six of the key variables that have been found to support well-being: income, healthy life expectancy, social support, freedom, trust, and generosity.

As a student, how happy are you? High school can be challenging—students are under a lot of stress from the amount of homework, studying for exams, preparing college applications, social anxieties, and athletic competitiveness. Teens routinely say their school-year stress levels are far higher than they think is healthy and their average reported stress exceeds that of adults, according to an annual survey published by the American Psychological Association.

Do you feel stressed because of the amount of homework you have?

The Census at School website contains a large database of responses to questions and responses from students in the United States and other countries.

Note: See the appendix for more information concerning the Census at School website (*ww2.amstat.org/censusatschool/students.cfm*).

Table 18.1

	Some or A Lot	Little or None	Total
US Teens			100
New Zealand Teens			100
Total	102	98	200

Sample result:

	Some or A Lot	Little or None	Total
US Teens	52	48	100
New Zealand Teens	50	50	100
Total	102	98	200

One question from the Census at School questionnaire students gave responses to is:

"How much pressure do you feel because of the schoolwork you have to do?"

□None □Very little □Some □A lot

A random sample from the Census at School database of 100 US 16- 17-year-old students who answered the question about stress found that approximately 54% of the students responded "some" or "a lot" of pressure.

A random sample from the New Zealand Census at School database of 100 New Zealand 16- 17-year-old students who answered the question about stress found approximately 48% of the students responded "some" or "a lot" of pressure.

Formulate a Statistical Question

The results of the two surveys indicate the two proportions of students who responded and feel "some" or "a lot" of pressure are different: 0.54 for U.S. 16- 17-year-old students and 0.48 for New Zealand 16- 17-year-old students. The difference between the two proportions is 0.06.

Are the two groups of students really not that far apart, or is the difference of 0.06 a significant difference? By significant difference, we mean the difference in the response proportions for the two samples is larger than what we would expect to see due to sampling variability.

The statistical question we want to answer is: "Is there a significant difference between the proportion of US 16- 17-year-old students and the proportion of New Zealand 16- 17-year-old students who responded they feel 'some' or 'a lot' of pressure because of schoolwork?"

Collect Appropriate Data

A simulation (similar to the one designed in Investigation 11) will be used to help answer the statistical question. In this simulation, a population is created of 200 students representing the combined number of US and New Zealand students who answered the question.

Create 200 slips of paper (Student Worksheet 18.3 Template) of the same size and mark 102 of the slips with an **L** to represent the 102 students (54 US and 48 New Zealand) who responded they feel "some" or "a lot" of pressure. Cut out the slips and thoroughly mix them. Then, randomly select 100 slips from the bag. These 100 slips represent the number of US teens. Count the number of slips with an **L** and record in the cell labeled US teens and "some" or "a lot." Based on this count, complete Table 18.1.

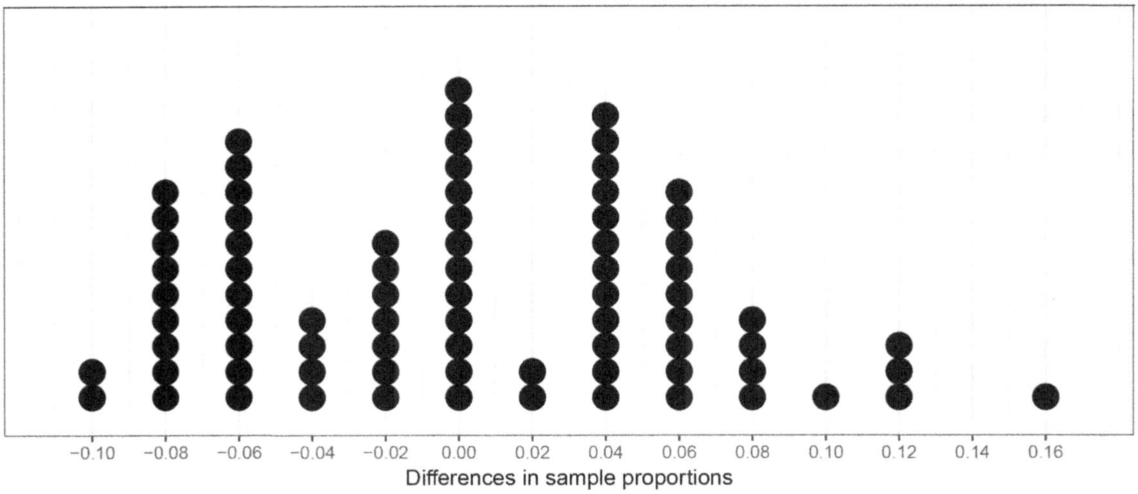

Figure 18.1: Dot plot showing sample result of 80 trials

Find the proportion of US teens who responded with "some" or "a lot" of pressure.

***Sample answer:** 52/100 or 0.52*

Find the proportion of New Zealand teens who responded with "some" or "a lot" of pressure.

***Sample answer:** 50/100 or 0.50*

Find the difference between these two proportions and record this difference.

***Sample answer:** 0.52-0.50 = 0.02*

Repeat this simulation a large number of times. Each time, record the difference between the two simulated proportions.

Analyze the Data

Construct a dot plot of the simulated differences between the two proportions.

A sample result (Figure 18.1) is based on 80 trials.

Interpret the Results in the Context of the Original Question

The original surveys showed 54% of US 16- 17-year-old students responded that they felt "some" or "a lot" of pressure and 48% of the New Zealand 16- 17-year-old students responded they felt "some" or "a lot" of pressure. This gave an observed difference of 0.06.

Based on the simulated differences, which were based on assuming there was no difference between the proportions, write a few sentences that address the statistical question: "Is there a significant difference between the proportion of US 16- 17-year-old students and the proportion of New Zealand 16- 17-year-old students who responded they feel "some" or "a lot" of pressure because of schoolwork?"

***Possible answer:** Based on the dot plot of the simulated differences, an observed difference of 0.06 would not be that unusual. Eighteen of the 80 (22.5%) simulated differences were greater than or equal to the observed difference of 0.06. This difference could be due to sampling variability and therefore there is no significant difference between the proportion of US 16- 17year-old students and the proportion of New Zealand 16- 17-year-old students who responded they feel "some" or "a lot" of pressure because of schoolwork. See Figure 18.2.*

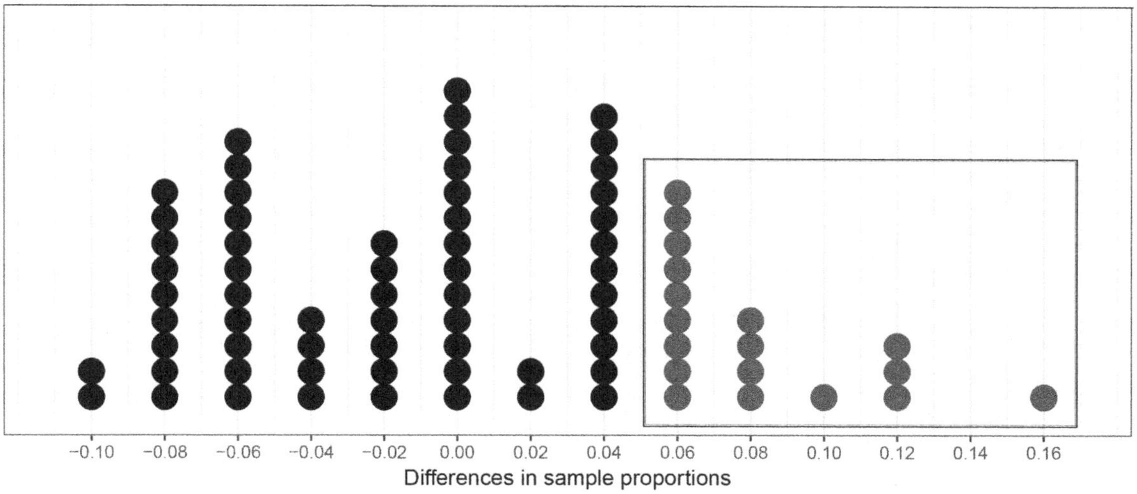

Figure 18.2: Dot plot of simulated differences

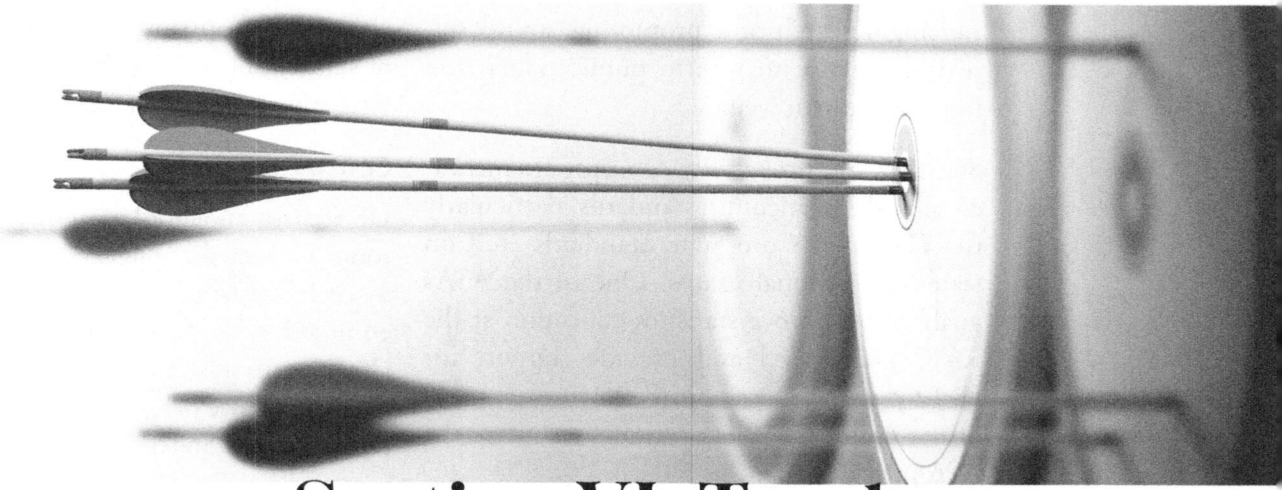

Section VI: Teacher Resources

The American Statistical Association (ASA) is the world's largest community of statisticians. The ASA supports excellence in the development, application, and dissemination of statistical science through meetings, publications, membership services, education, accreditation, and advocacy. Members serve in industry, government, and academia in more than 90 countries, advancing research and promoting sound statistical practice to inform public policy and improve human welfare.

Statistics and probability concepts are included in K–12 curriculum standards, particularly the Common Core State Standards, and on state and national exams. One of the ASA's goals is to improve statistics education at the K–12 grade level and provide support for K–12 classroom teachers. Following are some of the online K–12 educational resources the ASA provides.

For more information, visit *www.amstat.org/education.*

Online Resources

STatistics Education Web (STEW)

STatistics Education Web (STEW) is an online, searchable database of peer-reviewed lesson plans for K–12 teachers. Its content identifies both the statistical concepts being developed and the age range appropriate for its use. The statistical concepts follow the recommendations of the *Guidelines for Assessment and Instruction in Statistics Education.*

Teachers can navigate the site by grade level and statistical topic. For more information, visit *www.amstat.org/education/stew.*

Statistics Teacher

Statistics Teacher (*ST*) is an online journal published three times per year by the American Statistical Association/National Council of Teachers of Mathematics Joint Committee on Curriculum in Statistics and Probability for Grades K–12. *ST* is a free publication whose purpose is to keep K–12 teachers informed about statistical workshops; programs; and reviews of books, software, and calculators. In addition, articles include describing statistical activities that have been successful in the classroom. Contributors come from all levels of statistical expertise. For more information, visit *www.statisticsteacher.org.*

Census at School

US Census at School is an international classroom project that engages students in grades 4–12 in statistical problem solving. Students complete a brief online survey, analyze their class census results, and compare their class data with those of random samples of students in the United States and other countries.

This international program began in the United Kingdom in 2000 to promote statistical literacy in school children by using their own real data. The program is operative in the UK, New Zealand, Australia, Canada, South Africa, Ireland, Japan, and the United States. The US component of Census at School is hosted by the ASA's Education Outreach Program and cosponsored by partner Population Association of America.

The online survey asks students about topics such as the length of their right foot, height, favorite subject in school, and how long it takes them to get to school. Thirteen questions are common to every country participating in Census at School, but each country adds its own questions specific to the interests of its students. Periodically, the national data from the 13 common questions go to an international database maintained in the UK. For more information, visit *www.amstat.org/censusatschool.*

K–12 Statistics Education Webinars

The ASA offers free recorded web-based seminars on K–12 statistics education topics. This series was developed as part of the follow-up activities for Meeting Within a Meeting (MWM), a statistics workshop for math and science teachers held in conjunction with the Joint Statistical Meetings. For more information about the workshop, visit *www.amstat.org/asa/education/MWM/home*.

Some of the webinar topics available include the following:

- » A Statistician's Tour of the Common Core
- » Exploring Census at School Data with Fathom
- » What You Need to Know About the ASA Project Competition
- » Math Is Music: Statistics Is Literature
- » CSI Stats: Helping Students Become Data Detectives with the GAISE Framework
- » Doing Data Analysis in the Middle School with TinkerPlots
- » Working with K–12 Students to Create a Statistics Poster

For more information, visit *www.amstat.org/asa/education/K-12-Statistics-Education-ebinars.aspx*.

What's Going On in This Graph?

What's Going On in This Graph is a free, weekly online feature of the ASA and New York Times Learning Network. *New York Times* graphs of different types and context act as a springboard for middle- and high-school students in any course (college also welcome) to think critically about graphs. On most Wednesdays from September to April, graphs are released. Students respond to three questions: What do you notice? What do you wonder? What's going on in this graph? Teachers moderate their responses online from 9 a.m. – 2 p.m. ET. On Friday, the original article, additional questions, and "stat nuggets" —definitions of statistical terms and where they are seen in the graph—are revealed. No statistics background is necessary.

Publications

Bridging the Gap Between Common Core State Standards and Teaching Statistics

Bridging the Gap Between Common Core State Standards and Teaching Statistics includes 20 data analysis and probability investigations for teachers to use in their K–8 classrooms. Each investigation is based on the four-step statistical process as defined in the *Guidelines for Assessment and Instruction in Statistics Education (GAISE) Report: A Pre-K–12 Curriculum Framework.*

www.statisticsteacher.org/statistics-teacher-publications.

Statistical Education of Teachers

The *Statistical Education of Teachers (SET)* report outlines the content and conceptual understanding teachers require to assist their students in developing statistical reasoning skills. *SET* is intended for everyone involved in the statistical education of teachers, both the initial preparation of prospective teachers and the professional development of practicing teachers.

www.statisticsteacher.org/statistics-teacher-publications.

Making Sense of Statistical Studies

The *Making Sense of Statistical Studies* (*MSSS*) student module consists of 15 hands-on investigations that help students design and analyze statistical studies. It is written for an upper-middle-school or high-school audience having some background in exploratory data analysis and basic probability. The teacher's module includes supporting resources to help teachers use *MSSS*, as well as all the pages from the student module.

www.statisticsteacher.org/statistics-teacher-publications.

Data-Driven Mathematics

Data-Driven Mathematics is a series of modules funded by the National Science Foundation and written by statisticians and mathematics teachers. Intended to complement a modern mathematics curriculum in the secondary schools, the modules offer materials that integrate data analysis with topics typically taught in high-school mathematics courses and provide realistic, real-world data situations for developing mathematical knowledge. Scanned copies of these books are freely available to download (PDF).

» *Advanced Modeling and Matrices - Teacher's Edition*, by Gail Burrill, Jack Burrill, James Landwehr, and Jeffrey Witmer

  www.amstat.org/asa/files/pdfs/ddmseries/AdvancedModelingandMatrices--TeachersEdition.pdf

» *Advanced Modeling and Matrices*, by Gail Burrill, Jack Burrill, James Landwehr, and Jeffrey Witmer

  www.amstat.org/asa/files/pdfs/ddmseries/AdvancedModelingandMatrices.pdf

» *Exploring Centers - Teacher's Edition*, by Henry Kranendonk and Jeffrey Witmer

  www.amstat.org/asa/files/pdfs/ddmseries/ExploringCenters--TeachersEdition.pdf

» *Exploring Centers*, by Henry Kranendonk and Jeffrey Witmer

  www.amstat.org/asa/files/pdfs/ddmseries/ExploringCenters.pdf

» *Exploring Linear Relations - Teacher's Edition*, by Gail Burrill and Patrick Hopfensperger

  www.amstat.org/asa/files/pdfs/ddmseries/ExploringLinearRelations--TeachersEdition.pdf

» *Exploring Linear Relations*, by Henry Kranendonk and Jeffrey Witmer

  www.amstat.org/asa/files/pdfs/ddmseries/ExploringLinearRelations.pdf

» *Exploring Projects - Teacher's Edition*, by Emily Errthum, Maria Mastromatteo, Vince O'Connor, and Richard Scheaffer

  www.amstat.org/asa/files/pdfs/ddmseries/ExploringProjects--TeachersEdition.pdf

» *Exploring Projects*, by Emily Errthum, Maria Mastromatteo, Vince O'Connor, and Richard Scheaffer

  www.amstat.org/asa/files/pdfs/ddmseries/ExploringProjects.pdf

» *Exploring Regression - Teacher's Edition*, by Gail Burrill, Jack Burrill, Patrick Hopfensperger, and James Landwehr

  www.amstat.org/asa/files/pdfs/ddmseries/ExploringRegression--TeachersEdition.pdf

» *Exploring Regression*, by Gail Burrill, Jack Burrill, Patrick Hopfensperger, and James Landwehr

  www.amstat.org/asa/files/pdfs/ddmseries/ExploringRegression.pdf

» *Exploring Symbols - Teacher's Edition*, by Gail Burrill, Miriam Clifford, and Richard Scheaffer

www.amstat.org/asa/files/pdfs/ddmseries/ExploringSymbols--TeachersEdition.pdf

» *Exploring Symbols*, by Gail Burrill, Miriam Clifford, and Richard Scheaffer

www.amstat.org/asa/files/pdfs/ddmseries/ExploringSymbols.pdf

» *Exploring Systems of Inequalities - Teacher's Edition*, by Gail Burrill and Patrick Hopfensperger

www.amstat.org/asa/files/pdfs/ddmseries/ExploringSystemsofIneqalities--TeachersEdition.pdf

» *Exploring Systems of Inequalities*, by Gail Burrill and Patrick Hopfensperger

www.amstat.org/asa/files/pdfs/ddmseries/ExploringSystemsofIneqalities.pdf

» *Mathematics in a World of Data - Teacher's Edition*, by Jack Burrill, Miriam Clifford, Emily Errthum, Henry Kranendonk, Maria Mastromatteo, and Vince O'Connor

www.amstat.org/asa/files/pdfs/ddmseries/MathematicsinaWorldofData--TeachersEdition.pdf

» *Mathematics in a World of Data*, by Jack Burrill, Miriam Clifford, Emily Errthum, Henry Kranendonk, Maria Mastromatteo, and Vince O'Connor

www.amstat.org/asa/files/pdfs/ddmseries/MathematicsinaWorldofData.pdf

» *Modeling with Logarithms - Teacher's Edition*, by Jack Burrill, Miriam Clifford, and James Landwehr

www.amstat.org/asa/files/pdfs/ddm-series/ModelingwithLogarithms--TeachersEdition.pdf

» *Modeling with Logarithms*, by Jack Burrill, Miriam Clifford, and James Landwehr

www.amstat.org/asa/files/pdfs/ddmseries/ModelingwithLogarithms.pdf

» *Probability Models - Teacher's Edition*, by Patrick Hopfensperger, Henry Kranendonk, and Richard Scheaffer

www.amstat.org/asa/files/pdfs/ddmseries/ProbabilityModels--TeachersEdition.pdf

» *Probability Models*, by Patrick Hopfensperger, Henry Kranendonk, and Richard Scheaffer

www.amstat.org/asa/files/pdfs/ddmseries/ProbabilityModels.pdf

» *Probability Through Data - Teacher's Edition*, by Patrick Hopfensperger, Henry Kranendonk, and Richard Scheaffer

www.amstat.org/asa/files/pdfs/ddmseries/ProbabilityThroughData--TeachersEdition.pdf

» *Probability Through Data*, by Patrick Hopfensperger, Henry Kranendonk, and Richard Scheaffer

www.amstat.org/asa/files/pdfs/ddmseries/ProbatilityThroughData.pdf

Student Competitions

The ASA/NCTM Joint Committee on Curriculum in Statistics and Probability and the ASA's education department encourage students and their advisers to participate in its annual Data Visualization Poster Competition and Project Competition.

ASA Data Visualization Poster Competition for Grades K–12

A data visualization poster is a display containing two or more related graphics that summarize a set of data, look at the data from different points of view, and answer specific questions about the data.

www.amstat.org/asa/education/ASA-Statistics-Poster-Competition-for-Grades-K-12.aspx

ASA Statistics Project Competition for Grades 7–12

A statistical project is the process of answering a research question using statistical techniques and presenting the work in a written report.

www.amstat.org/asa/education/ASA-Statistics-Project-Competition-for-Grades-7-12.aspx

Made in the USA
Coppell, TX
05 November 2026

11331606R00122